LABORATORY INVESTIGATIONS IN CELL BIOLOGY

Allyn A. Bregman

Professor of Biology
State University of New York—College at New Paltz

LABORATORY INVESTIGATIONS IN CELL BIOLOGY

2nd EDITION

JOHN WILEY & SONS

New York • Chichester • Brisbane • Toronto • Singapore

COVER ILLUSTRATION: Densitometric tracing of an electropherogram of the LDH isozymes in a commercially prepared extract of rat heart. The LDH isozymes were separated using agarose gel electrophoresis, as described in Project 6.

ISBN 0-471-82569-7

Printed in the United States of America

10 9 8 7 6 5 4 3 2

CONTENTS

PREFACE

The second edition of *Laboratory Investigations in Cell Biology* consists of 18 projects for the laboratory in undergraduate cell biology courses. As in the first edition, the projects cover a broad range of cell phenomena, but there is now a greater emphasis on biochemistry. In addition, the topical sequence has been changed to parallel more closely the sequence found in most cell biology texts. The early projects (1–6) are concerned with cytochemistry and biochemistry. The next seven projects (7–13) focus on the major cytoplasmic organelles and fundamental cellular processes. Projects 14 through 17 examine the chromosomes and nuclear events. The last project (18) is devoted to photomicrography.

As the cell topics progress and build, so do the techniques. A technique or methodology is introduced, and then, in one or more subsequent projects, it is utilized or further developed. The methods presented in the manual include microscopy, staining reactions, spectrophotometry, electrophoresis, cell fractionation, chromatography, and photomicrography. The most significant change in the second edition is the inclusion of two projects on electrophoresis: the electrophoresis of hemoglobin (Project 5) and of LDH isozymes (Project 6). In the area of cytochemistry there are now two projects (2 and 3), each dealing with a single staining reaction. Cell fractionation methods are now presented in a separate project (8), and photomicrography is written in the form of a project incorporating instant 35mm film as one of the options.

The format of the individual projects remains the same. Each begins with an introduction, which is intended to be read before class. It provides the background theory that students need to understand the biological phenomenon being investigated, the methodology, and the procedural steps. In the procedure itself, the steps are spelled out so that students can make acceptable preparations without the need for constant directions by the instructor. The collection of data is facilitated by inclusion of data sheets, graph paper, and drawing paper. There are many new exercises and questions at the end of the projects to guide students in drawing conclusions from the laboratory observations. All exercises, questions, and data sheets are separate from the procedural steps and can be removed for grading purposes.

Another goal retained in the second edition is workability. Most of the projects can be carried out with the equipment and supplies commonly found in teaching laboratories—items like the compound microscope, spectrophotometer, incubator, and clinical centrifuge. Electrophoresis equipment is required for Projects 5 and 6, a refrigerated, high-speed centrifuge for part of Project 8 and for Project 9, and photomicrographic equipment for Project 18. The equipment, supplies, and reagents required for the individual projects are given in the Instructor's Guide, which follows Project 18. In four projects (2, 3, 16, 17), directions for pre-prepping the labs are included in the Instructor's Guide. Pre-prepping allows the bench work in Projects 2, 3, 16, and 17 to be completed in a single 3-hour laboratory period, as in the other projects. The pre-prepping also allows students to begin the procedures in these four projects with slides of fixed blood cells that are ready for staining.

I am indebted to a number of individuals for their contributions to the second edition. My wife, Sybil, has provided invaluable assistance, critically reading the manuscript at every stage. I am grateful to my department chairman, Arnold Nemerofsky, for his suggestions and encouragement as the electrophoresis projects were developed, and to David B. Straus (Department of Chemistry, SUNY, New Paltz) for his helpful comments on two projects. I would also like to acknowledge the helpful review of the manuscript by David J. Saxon (Department of Biological Sciences, Morehead State University, Morehead, Kentucky). Acknowledgment is also due the very able staff at Wiley, especially my editor, Katie Vignery, and production supervisor, Rita Kerrigan.

Allyn A. Bregman
New Paltz, N.Y.

LABORATORY INVESTIGATIONS IN CELL BIOLOGY

MICROSCOPY

INTRODUCTION

For most biologists, the microscope is an important tool, but for the cell biologist, it is indispensable! From the earliest observations of protozoans by Antony van Leeuwenhoek (1632–1723) to today's sophisticated research with fluorescence and interference optics, the microscope has remained the basic instrument for probing the structural basis of life. The microscopes of van Leeuwenhoek are considered *simple microscopes*, for there is but a single lens, as in a magnifying glass. Much greater magnification is achieved with a two-lens system, a *compound microscope*. In this project, you will learn the proper use of the compound microscope as you examine some plant cells, a protist, and bacteria.

THE COMPOUND MICROSCOPE

The construction of a modern bright-field microscope is shown in Figure 1.1. The *illuminator* in most modern microscopes is built into the base. It consists of a high-intensity bulb, transformer, lens system, and adjustable *field diaphragm*, or *field iris*, that controls the size of the illuminated field of view. Light passes from the illuminator window to the *substage condenser*, an adjustable lens system that focuses the light on the object. In the lower part of the substage condenser there is an *iris diaphragm*, which controls the diameter of the light beam entering the condenser. The two lens systems that are involved in image formation are the *objective*, which is close to the object, and the *ocular*, or *eyepiece*, which is close to the eye. The objective produces an enlarged and inverted projection of the object on the other side of the lens. This first image (a real image) serves as the object for the ocular. The ocular produces a final image (a virtual image) that is greatly enlarged and still inverted.

Magnification.   The magnifications in a typical set of objectives are 10× (low power), 40× (high-dry), and 100× (oil-immersion). Most eyepieces have a magnification between 8× and 12.5×. The *total magnification* is the product of the magnifications of the objective and ocular. Thus, the magnification with a 100× objective and 10× ocular is 1000×.

Resolution and Numerical Aperture.   In addition to magnifying the image, the microscope resolves detail, i.e., reveals the subunits of which an object is composed. The resolving power of a microscope is specified by the *limit of resolution (l.r.)*, defined as the smallest distance by which two neighboring points can be separated and still be discerned as separate entities. The l.r. of the naked eye is about 0.1 mm, whereas that of a high-quality light microscope is about 0.2 μm, a 500-fold improvement. Thus, if two dotlike structures in a microscope specimen are separated by less than 0.2 μm, they will appear as a single, oblong dot.

Optical and Mechanical Features of
THE MICROSCOPE

Retinal Image

Eyepoint

Wide Field Eyepiece

Real Image

Arm

Focusable Rotatable Nosepiece

Infinity Corrected Objectives

Slide Holder Assembly

Condenser Focusing Knob

Coarse Adjustment Knob

Fine Adjustment Knob

Field Diaphragm

Lamp Filament

Virtual Image Distance 250 mm

Inclined Reversible Body

Mechanical Stage

Virtual Image

Iris Diaphragm Lever

Substage Condenser

Illuminator Window

Illuminator Controls

In-Base Illuminator and Transformer

Base

FIGURE 1.1
Components and optical path in a modern microscope. (Courtesy of Reichert Scientific Instruments.)

The limit of resolution for a microscope is calculated from the *Abbe equation*,

$$\text{l.r.} = \frac{0.61\lambda}{\text{N.A.}}$$

where λ is the wavelength of the light used to view the object and *N.A.* is the *numerical aperture*, a measure of the cone angle of light entering the objective lens (Fig. 1.2). As can be seen from the Abbe equation, the limit of resolution decreases as the N.A. increases. Thus, the larger the N.A., the greater is the resolving power. The increased resolving power results from the larger numerical aperture reducing the *diffraction pattern*, which is the fringe of dark and light rings around object borders. When the fringes become too great, it becomes impossible to discern adjacent objects as separate entities. The highest-quality objectives have an N.A. of 1.40. The lowest possible limit of resolution attainable with a light microscope accrues when a 1.40-N.A. objective is used with a violet filter ($\lambda = 400$ nm). According to the Abbe equation, the limit of resolution equals 0.61(400 nm)/1.40, or 0.17 μm.

In practice, however, the microscope may be operating at a lower N.A., resulting in a somewhat larger l.r., about 0.25–0.40 μm. The *working N.A.* of the microscope equals *n sine* θ. The angle θ is one-half the cone angle of light entering the front lens of

FIGURE 1.2
Light path from the substage condenser to the front lens of the objective.

the objective (Fig. 1.2). The cone angle is controlled by the iris diaphragm of the substage condenser. Opening the *condenser iris* increases the cone angle of light. The condenser iris must be sufficiently open for the working N.A. to equal the N.A. engraved on the barrel of the objective, which is the highest value attainable for that lens. Usually, the condenser iris is partially closed to improve image contrast. When this adjustment is made, the cone angle and, hence, the working N.A. are reduced. Thus, an objective that has an N.A. of 1.25 engraved on the barrel is usually operating at an N.A. that is close to 1.0. The working N.A. is also a function of n, the refractive index of the medium between the coverslip and the objective lens. With the low power and high-dry objectives, the medium is air ($n = 1.0$), but with the oil-immersion objective, a drop of *immersion oil* ($n = 1.515$) is placed between the coverslip and the lens. The refractive index of immersion oil is approximately the same as that for glass so that light is not refracted as it passes from the object to the objective.

Lens Characteristics.   All lenses have aberrations. Two types of aberrations that are of particular concern in microscopy are *chromatic* and *spherical*. Left uncorrected, chromatic aberration results in color fringes around objects. When spherical aberration is present, the image may be fuzzy, and straight lines may appear curved. Your microscope is probably equipped with *achromat objectives*, which are corrected chromatically for two colors (blue and red) and spherically for one (green). Another important lens correction is for *curvature of field*. With an uncorrected objective, the entire field is not in focus at the same focal setting. Objectives designed to produce a *flat field* are designated with the prefix *plan*.

Several significant optical characteristics of a microscope are determined by the N.A. and the magnification of its lenses. These are summarized in Table 1.1 for a typical set of planachromat objectives. The *maximum useful magnification* is the limit that increased magnification can improve our ability to resolve detail. Any greater magnification will not improve our ability to discern but only reduce the quality of the

TABLE 1.1
Relationship of N.A. and objective magnification to maximum useful magnification, working distance, depth of field, and diameter of field. Data are for Reichert planachromat objectives used with 10× wide field oculars in a Series 110 microscope. (Data courtesy of Reichert Scientific Instruments.)

Objective lens	N.A.	Maximum useful magnification (1000 × N.A.)	Working distance	Depth of field	Diameter of field
Plan Achro 10×	0.25	250×	9.1 mm	8.8 μm	2.0 mm
Plan Achro 20×	0.50	500×	1.4 mm	2.2 μm	1.0 mm
Plan Achro 40×	0.66	660×	0.5 mm	1.3 μm	0.5 mm
Plan Achro 100× (oil)	1.25	1250×	0.1 mm	0.5 μm	0.2 mm

final image. Depending on lens quality, the maximum useful magnification can be as great as 1000 times the N.A. Another characteristic of the objective is its *working distance*, defined as the clearance between the top of the coverslip and the lowest part of the objective when the specimen is in focus. As the N.A. increases, the working distance decreases dramatically in order to accommodate the larger cone angle of light. The *depth of field*, which is the specimen thickness that is in focus at one focal setting, also decreases sharply with increasing N.A. Finally, there is the *diameter of the field*, which is inversely proportional to the total magnification; with a doubling of magnification, the diameter of the field is halved.

Köhler Illumination.   To achieve the highest intensity and most uniform illumination from a nonuniform source, such as the filaments of a bulb, the microscope components must be adjusted for *Köhler illumination*. The two basic steps in the method, which was introduced by August Köhler in 1893, are (1) focusing the bulb filaments at the plane of the condenser iris and (2) focusing an image of the field iris at the plane of the object. In microscopes with an in-base illuminator, the illuminator lens system is positioned at the factory so that step 1 need not be performed. You will carry out step 2 by adjusting the height of the substage condenser. After obtaining a bright and uniformly illuminated field, you will adjust the aperture of the condenser iris to obtain the best compromise of resolving power and contrast.

Contrast, Brightness, and Filters.   Contrast is controlled by the condenser iris. Generally, the condenser iris should be open so that about two-thirds the diameter of the back focal plane of the objective is filled with light. With a lightly stained or unstained specimen, the condenser iris may have to be further closed; resolving power will be reduced, but contrast and depth of field will be dramatically improved. Image contrast is also enhanced by using an appropriate colored filter. In the calculation of the lowest possible l.r., it was assumed that a violet filter was used, but, in practice, a *green* filter (λ = 500–600 nm) is placed in the light path. The green illumination is nearly complementary to the red and purple hues produced by many biological stains and, thereby, increases contrast. In addition, the human eye is most sensitive to green light, and objectives are corrected for aberrations in this spectral region.

Brightness is controlled by the transformer setting and by *neutral density filters*, which reduce light intensity without altering the color quality of the illumination. The brightness should never be reduced by closing the condenser iris or by lowering the substage condenser since there will be a concomitant loss in resolution. Bulb life can be prolonged by using the lowest transformer setting and by not switching the illuminator on and off repeatedly. The illuminator should be left on until all observations have been made for that laboratory session.

Microscope Technique.   All modern microscopes are *parfocal* and *parcentral*. The term parfocal indicates that the focus does not change appreciably when the objective is changed; only slight touching up of the focus with the fine adjustment knob is required. The term parcentral indicates that the center of the field does not change when the objective is changed. Before the objective is changed, the object should be centered in the field.

Always start focusing with low power first, checking that the coverslip is on the upper surface of the slide so that the higher-power objectives will not hit the slide. To obtain a good image with the high-dry objective, a coverslip must be used, and it should be of No. 1½ thickness (0.17–0.18 mm). Oil-immersion objectives are so designated and *must* be used with immersion oil between the front lens and the coverslip. The oil-immersion objective is not as sensitive to coverslip thickness and can even be used without a coverslip. Keep in mind that the oil-immersion objective has a very short working distance and can easily impact on the slide. When the objective is switched back to high-dry, it is necessary to wipe the immersion oil from the coverslip. After all observations have been made, the oil-immersion lens should be wiped with clean lens paper. *Wide field eyepieces* are a great advantage since they produce a large field of view and place the eyepoint (point where the exiting light rays converge) far enough from the lens to provide adequate clearance for eyeglasses. Actually, since the microscope's optics can correct for nearsightedness and farsightedness, eyeglasses need not be worn unless there is severe astigmatism. So that your eyes are at the level of the eyepieces and viewing is comfortable, it is important that the chair height be properly adjusted.

Image quality is dramatically impaired by dust and body oils on the lenses. The ocular can be cleaned with lens paper moistened with distilled water. When the microscope is not in use, it should be covered to keep dust off the lenses. Finally, the microscope should always be carried in an upright position, with two hands.

PHASE-CONTRAST MICROSCOPY

When detailed study of unstained specimens is required, we generally rely on the phase-contrast microscope. With phase optics, details of a low-contrast object are made apparent because of a phenomenon known as *interference*. If two light waves arrive at some point with their crests and troughs in phase, there is *constructive interference* and a great increase in brightness. Conversely, if the two waves are completely out of phase, there is *destructive interference*; the two waves cancel each other out, and the intensity is zero. When light waves pass through a typical unstained cell preparation, they are slowed, so that as they emerge from the specimen they are retarded by about one-fourth of a wavelength ($\frac{1}{4}\lambda$). The special light path in a phase-contrast microscope (Fig. 1.3) selectively retards these light waves yet another $\frac{1}{4}\lambda$. Compared to the light waves that did not strike the object, they are now out of phase by $\frac{1}{2}\lambda$ (inset at upper left of Fig. 1.3). The visual effect of the $\frac{1}{2}\lambda$ interference is a dark object contrasted against a lighter background.

The contrast enhancement in a phase-contrast microscope is effected by a special substage condenser and phase objectives. The substage condenser contains a rotatable turret mount with several *annular diaphragms* (bottom of Fig. 1.3), one corresponding to each phase objective. One position has no annulus so that bright-field illumination can be obtained. Each annulus directs a hollow cone of light to the specimen. The light waves that do not strike the object are termed the *direct rays*, for they are not diffracted and continue on path to the objective. Those light waves that do strike the object are retarded by $\frac{1}{4}\lambda$ and are diffracted from their path. At the back of the phase objective there is a *phase plate*, containing a ring-shaped disk that accommodates the direct rays and allows them to pass through without retardation. The ring also contains absorbing material to reduce the intensity of the direct rays. The remainder of the phase plate is composed of material that retards light waves by $\frac{1}{4}\lambda$. The *diffracted rays*, which strike all regions of the phase plate, are thereby retarded by a total of $\frac{1}{2}\lambda$ ($\frac{1}{4}\lambda$ by the specimen plus $\frac{1}{4}\lambda$ by the phase plate). To achieve maximum contrast, each annulus must be aligned with its respective phase plate. A *phase telescope* or built-in *telescopic lens* is used for viewing both the annulus and phase plate during the alignment procedure.

PROCEDURES

FAMILIARIZATION WITH MICROSCOPE COMPONENTS

1. Familiarize yourself with your microscope by identifying the components labeled in Figure 1.1. Trace the light path from the illuminator to the eyepiece.
2. Note the magnification of each objective and of the ocular on your microscope. Then compute the *total magnification* with each objective and enter the values in the table on page 11.

BRIEF DESCRIPTION OF PHASE

Essentially, phase contrast is produced by the combination of (a) an annular diaphragm located below the condenser that directs a hollow cone of light through the specimen and (b) a phase plate at the back focal plane of the objective. Some of the light passing through the transparent specimen is diffracted by slight differences in optical path (refractive index × thickness) and moves so as to be distributed over the whole aperture of the objective. The balance of the light passes directly through the specimen as a cone of concentrated light toward coincidence with the "ring" of the phase plate. The phase plate alters the intensity and phase relationships of the diffracted and direct light so that when they recombine to form an image, invisible specimen optical path differences are converted into visible light-intensity differences.

FIGURE 1.3
Optical path in a phase-contrast microscope. (Courtesy of Reichert Scientific Instruments.)

3. Note the N.A. of each objective and compute the best limit of resolution attainable with each. Assume that a green filter (λ = 550 nm) is used and that the working N.A. with the oil-immersion objective is 1.0. Enter these values in the table on page 11.
4. Determine the maximum useful magnification and the highest-power ocular to be used for your microscope. Enter both values below the table on page 11.

OBTAINING KÖHLER ILLUMINATION WITH AN IN-BASE ILLUMINATOR

1. Turn on the illuminator, using the lowest transformer setting.
2. By turning the condenser focusing knob, raise the substage condenser to the top of its travel. Open the field iris and the condenser iris all the way.
3. Place a slide with a stained specimen on the stage, checking that the coverslip is on the upper surface. Position the slide so that the specimen is in the light path.
4. Using the 10× objective, bring the object into focus. Start with the objective close to the slide and, with the coarse adjustment knob, increase the distance between the objective and the slide until the object is in focus. Touch up the focus with the fine adjustment knob. If the field is uncomfortably bright, place a neutral density filter over the illuminator window.
5. In a binocular microscope, an accommodation must be made for differences in interpupilary distance; move the ocular tubes closer together or farther apart until both eyes see the image simultaneously. To adjust for differences in the vision of the two eyes, bring the object into sharp focus with the fine adjustment knob while looking through the eyepiece in the fixed-length ocular tube, which is usually the right one. The left eye should be closed or covered with a piece of paper. Now, with the right eye closed or covered, focus the image for the left eye by rotating and, thereby, changing the length of the left ocular tube.
6. Close the field iris almost all the way. Then slightly lower the height of the substage condenser until there is a sharp image of the inner leaves of the field iris superimposed on the image of the object.
7. Center the image of the field iris with the centering screws on the illuminator housing or, if it is not adjustable, with the centering screws on the substage condenser mount. After it is centered, open the field iris until the light just completely fills the field of view. The result of steps 6 and 7 is a uniformly illuminated field.
8. Remove an eyepiece and look down the ocular tube. Close the condenser iris until about two-thirds the diameter of the back focal plane of the objective is filled with light.
9. Replace the eyepiece. The microscope is now properly adjusted for the 10× objective.
10. Change to the high-dry objective and touch up the focus with the fine adjustment knob. Adjust both iris diaphragms as follows. Close the field iris until the light just completely fills the field of view and, if necessary, touch up the focus and centering of the field iris. Open the condenser iris until about two-thirds the diameter of the back focal plane of the objective, observed through an empty ocular tube, is filled with light. Replace the eypiece. The microscope is now properly adjusted for the high-dry objective.
11. Partially rotate the nosepiece, place a drop of immersion oil on the coverslip, and then bring the oil-immersion objective into place. Touch up the focus with the fine adjustment knob. Close the field iris until the light just completely fills the field of view. Generally, the field is so small with the oil-immersion objective that the field iris will be closed to its smallest aperture. The field iris need not be focused if this was previously done for the high-dry objective. Open the condenser iris until about two-thirds the diameter of the back focal plane of the

objective, observed through an empty ocular tube, is filled with light. Replace the eyepiece. The microscope is now properly adjusted for the oil-immersion objective. For routine visual observations, the eyepiece need not be removed to adjust the condenser iris every time the objective is changed; one simply adjusts it until there is good contrast.

CALIBRATION OF AN OCULAR MICROMETER

The ocular of a microscope can accommodate a glass disk with markings (pointer, cross hairs, etc.) that are in focus with the object. For object measurement, use an *ocular micrometer*, a disk with calibrated divisions.

1. With the ocular micrometer in place, focus on the calibrations of a stage micrometer slide using the low power objective.
2. Determine the length of each division (in micrometers) by dividing the total length subtended on the stage scale by the total number of divisions on the eyepiece scale. Enter the values in the table on page 11.
3. Carefully change to the high-dry objective and determine the length (in micrometers) of each division. Enter the values in the table on page 11.
4. Do *not* attempt to focus on the calibrations with the oil-immersion objective, which could easily impact on the thick micrometer slide. Instead, calculate the length of each division from the value obtained at low power. Divide the value for micrometers per division by 10 (assuming a $100 \times$ oil-immersion objective and a $10 \times$ low power objective, i.e., $100/10 = 10$). Enter the value in the table on page 11.

WET MOUNTS

In a wet mount, a piece of tissue is placed on a clean slide with a drop of water, stain, or reagent. A coverslip is then gently lowered on the preparation.

Onion Epidermis.
1. Add a small drop of water to the center of a clean slide.
2. Cut a fresh onion in half and remove a layer.
3. Using pointed forceps, strip a small piece of epidermis from the concave surface of a layer and place it on the drop of water, being careful that it does not fold over on itself. Add a drop of water and a coverslip.
4. Examine the cells with the low power and high-dry objectives. Answer the first four questions on page 11.
5. Add one or two drops of *Lugol's iodine solution* at one edge of the coverslip and draw it through by touching a piece of paper towel to the opposite edge of the coverslip. Examine the stained cells (at the edge where the stain was added) with the low power and high-dry objectives, varying the opening of the condenser iris until the nucleus is clearly visible. How does the adjustment of the condenser iris differ from that used for the unstained cells? Answer on page 11.
6. Measure the diameter of 10 nuclei, selecting cells in which the nucleus appears round. Calculate the *mean* ($\overline{X}$), defined as $\overline{X} = \Sigma X/N$, where ΣX is the sum of the individual values and N is the total number of measurements. Enter the 10 values and the mean on page 12.
7. On the blank paper provided at the end of this project (p. 13), draw a typical onion epidermal cell, labeling the visible cell structures. Every drawing done should have a legend containing the name of the organism (here, *Allium cepa*), the cell type, the stain or technique used in its preparation, and the magnification of the drawing, or *plate magnification*, which is (size of object in the

drawing)/(actual size of object). For example, if the length of a cell in a drawing is 70 mm and its actual length is 35 μm, the plate magnification is (70 mm/35 μm) = 2000×. Don't make the mistake of recording the magnification at which the object was viewed!

Elodea Leaf.

1. Using a healthy sprig of *Elodea*, prepare a wet mount of a leaf in a drop of water. The upper surface of the leaf should be facing up.
2. Before focusing on the preparation and with the *low power* objective in place, turn the coarse and fine adjustment knobs in one direction and then in the other direction as you watch the movement of the nosepiece or stage from the side. Determine whether the focal plane moves up or down when you turn the focus knobs away from you and toward you.
3. With the green filter removed, examine the cells under low power and high-dry. Determine how many cell layers there are and which layer is composed of larger cells. Enter your answers on page 12.
4. Using the high-dry objective, determine the thickness of the *Elodea* leaf as follows. Record the line division on the fine adjustment knob when the upper leaf surface just comes into focus and again when the lower leaf surface just starts to go out of focus. The number of divisions multiplied by the distance/division (usually 1 μm/division) gives the vertical distance traversed. Repeat twice, recording the three values and the mean on page 12.
5. Scan the preparation for cells exhibiting *cyclosis*, or *cytoplasmic streaming*, and determine the approximate rate in one cell. To obtain the rate of cyclosis, measure the length of the cell and the time required for a chloroplast to move from one end of the cell to the other. Enter the data on page 12.
6. Draw a typical *Elodea* leaf cell, labeling the cell wall, cytoplasm, vacuole, chloroplasts, and nucleus (if visible). Record the plate magnification.

Euglena.

1. On a clean slide, place a drop of the *Euglena* culture and add a coverslip. Observe several specimens as they swim across the field. Look closely for all the ways in which they are moving and then answer the question on page 12. Locate a nonswimming specimen and observe its characteristic movement, termed *euglenoid movement*. Describe euglenoid movement on page 12.
2. Prepare a fresh wet mount using one drop of the *Euglena* culture and a small drop of *Protoslo* (Carolina Biological Supply Co.), a viscous liquid used to slow swimming protozoa. Mix with a toothpick, add a coverslip, and examine under low power and high-dry. Identify the anterior (leading) and posterior (trailing) ends of the *Euglena* and determine from which end the flagellum emerges. Remember to close the condenser iris to enhance contrast. Enter your answer on page 12.
3. Examine the internal structure of the slowed *Euglena*. Draw a *Euglena*, labeling all visible structures and recording the plate magnification.

PREPARED SLIDES

Zea Stem, Cross Section.

1. Scan the stained preparation under low power and identify the numerous *vascular bundles* scattered near the periphery of the stem.
2. Center a vascular bundle in the field, add a drop of immersion oil, and switch to the 100× objective. Identify the large *xylem vessel elements*, the cell walls of which are stained red.
3. With the green filter in place, measure the thickness of the cell wall in three

vessel elements (in regions that are not in contact with the wall of another vessel element). Enter the values and the mean on page 12.

4. Examine the preparation without the green filter and compare the image to that observed *with* a green filter. Explain on page 12.

Bacteria.

1. Scan a stained preparation under high-dry to locate an area with a concentration of bacteria.
2. Switch to oil-immersion and examine the morphology of the cells.
3. Draw a representative cell or group of cells for each bacterial species supplied. Next to each drawing, give the dimensions and plate magnification.

OPTIONAL PROCEDURE

PHASE-CONTRAST MICROSCOPY

1. Turn on the illuminator. An intense light source is required for phase microscopy, and colored filters are not used for visual observations.
2. Rotate the turret on the phase condenser to the "open" position, i.e., to the position without any annular diaphragm.
3. With a stained specimen on the stage and the 10× phase objective in place, adjust the microscope for Köhler illumination (through step 7, p. 7).
4. Open the iris diaphragm of the substage condenser; the condenser iris is left fully open for *all* phase microscopy.
5. Rotate the turret to bring the 10× annulus into position.
6. Replace the right eyepiece with a phase telescope and adjust its length until the annulus and phase plate are in sharp focus. In some microscopes, a telescopic lens is built into the body and must be swung into position.
7. While looking through the phase telescope, turn the centering wrenches in the turret and align the 10× annulus so that it is concentric with the ring in the phase plate of the objective.
8. Center the other annuli for their respective objectives. When changing phase objectives, check that the field iris is centered and that its image is focused at the same plane as the object; if the condenser height is incorrectly adjusted, the annulus image will be partially colored. Replace the phase telescope with the regular eyepiece.
9. Place a wet mount of unstained onion epidermis and/or *Euglena* on the stage and examine with the low power and high-dry phase objectives. What structures are more clearly delineated with phase than with bright-field optics? Enter your answers on page 12.

REFERENCES

Curry, A., Grayson, R. F., and Kosey, G. R. 1982. *Under the Microscope.* Van Nostrand Reinhold, New York.

Locquin, M. and Langeron, M. 1983. *Handbook of Microscopy.* Butterworths, London.

Spencer, M. 1982. *Fundamentals of Light Microscopy.* Cambridge Univ. Press, Cambridge, England.

OBSERVATIONS AND QUESTIONS

FAMILIARIZATION WITH MICROSCOPE COMPONENTS

	Magnification	Total magnification	N.A.	l.r.
Low Power	10	100	.25	1.342 μm
High-dry	40	400	.65	.516 μm
Oil-immersion	100	1000	1.25	.268 μm
Ocular	10	—	—	—

Maximum useful magnification __1000__

Highest-power ocular to be used __10__

CALIBRATION OF AN OCULAR MICROMETER

	Length subtended on stage scale (mm)	Length, μm/division
Low power	1 BMU .1mm	100 × 100 10,000
High-dry	4 OMU .1mm	25 × 400 10,000
Oil-immersion		10 × 1000 10,000

WET MOUNTS

Onion Epidermis.
How many cell layers are there? __one__

What cell parts are visible? ___none – plasma membrane___
How must the condenser iris be adjusted to observe the nucleus?

_____must be shut_____

Why? _____
How has the adjustment of the condenser iris changed when observing the stained preparation?

_____

Nuclear diameters (μm)

_25_ _30_ _20_ _35_ _14_ _25_ _31_ _27_ _25_ _20_

$\bar{X} = $ _25.2_ μm

Elodea Leaf.

Number of cell layers _2_

Layer with larger cells _bottom_

Leaf thickness (μm) _17_ _8_ _10_ $\bar{X} = $ _11.67_ μm

Length of cell = _75_ μm; time required for movement = _62_ sec

Rate of cyclosis = _6.25_ μm/sec   .0375 cm/hr

Euglena.

What other motion do swimming specimens exhibit besides forward motion?

jig - jag   reverse

Describe the motion of nonswimming *Euglena*, i.e., euglenoid movement.

round and rand - quiver

Does the flagellum emerge from the anterior or posterior end of the organism?

anterior

PREPARED SLIDES

Zea Stem, Cross Section.
Cell wall thickness (μm) _12_ _7_ _5_ $\bar{X} = $ _8_ μm
How has the image improved with a green filter?

little more clear

PHASE-CONTRAST OBSERVATIONS

What structures are more apparent in onion epidermis?

What structures are more apparent in *Euglena*?

500 X

onion cell

$$\frac{75\ \mu m}{12\ sec} \times \frac{10^{-4}\ cm}{1\ \mu m} \times \frac{60\ sec}{hr}$$

400 X

2000 X

25 mm

Euglena

coccus
2 mm

5000 X

~~3 mm~~
2.25 mm

6700 X

10 mm

4500 X

CELLULAR CARBOHYDRATES

INTRODUCTION

Carbohydrates and macromolecules that contain sugar moieties perform a variety of functions in the cell and in the organism. The glycogen in animals and the starch in plants serve as energy stores. Other carbohydrates, such as chitin and cellulose, have structural roles in animals and plants, respectively. Oligosaccharides are components of the glycoproteins and glycolipids found in the plasma membrane. These sugar chains, which always face the cell's exterior, stabilize the position of glycoproteins and glycolipids within the membrane, function in cell adhesion, and confer immunological specificity to the cell surface.

Every class of macromolecule can be localized in cells with specific cytochemical reactions. The term *cytochemistry* can refer to any methodology that probes the chemical nature of the cell but usually is reserved for specific staining reactions and subsequent microscopic analysis. In this project, you will use a cytochemical method, specifically, the *periodic acid–Schiff (PAS) reaction*, to identify cellular carbohydrates in human blood cells. The PAS reaction stains all insoluble polysaccharides (glycogen, starch, cellulose), mucopolysaccharides, mucoproteins, glycoproteins, and glycolipids. After staining the blood smears, you will be able to localize PAS-positive material in white blood cells and in other formed elements of the blood. You will first learn to identify the blood cell types by examining a prepared slide of a stained blood smear.

BLOOD CELL TYPES

Histologists frequently use *Wright–Giemsa stain* for the identification and study of human blood cells. Wright–Giemsa stain is a mixture of methylene blue, methylene azure, and the eosinates of both. The azures act as bases and stain the *basophilic* (base-loving) elements of the cells blue, while the eosins behave as acids and stain the *acidophilic* (acid-loving) structures red. Since there is a combination of dyes and a varying affinity for each, the various parts of the cell are stained in hues of pink, purple, blue, and red. After Wright–Giemsa staining at pH 6–7, the structures in blood cells have characteristic colors, and each cell type is readily identifiable.

Figure 2.1 is a diagram of the various human blood cell types as they appear in a fixed smear. There are two classes of cells, red blood cells, or *erythrocytes*, and white blood cells, or *leukocytes*. The leukocytes are placed into two groups, the *granulocytes*, which have conspicuous cytoplasmic granules, and the *argranulocytes*, which lack them. The granulocytes include the *neutrophils*, *eosinophils*, and *basophils*. The agranulocytes include the *lymphocytes* and *monocytes*. In addition to these cell types there are the *platelets*, which are small cell fragments.

Erthrocytes. The erythrocyte is the most frequent cell type. It is also the smallest blood cell, about 7.5 μm in diameter and 1.9 μm in thickness at the periphery. Human erythrocytes lack a nucleus and are biconcave. In Wright–Giemsa–stained preparations, they appear pink with a central region staining lighter because of the concavity. They lack any internal organelles but are filled with hemoglobin for the function of oxygen transport.

Neutrophils. Neutrophils are the most frequent type of white blood cell, comprising 55–75% of the leukocyte population. In stained fixed smears, neutrophils are round, 9–12 μm in diameter, and easily recognized by their dark purple, multilobed nucleus. Most are *segmented neutrophils*, in which the nucleus has 3–5 lobes connected by a thin strand of chromatin. The young neutrophils, which do not have their nuclei divided into lobes, are called *band neutrophils*. Because of the varied nuclear morphology, neutrophils are also termed *polymorphonuclear leukocytes*. The cytoplasm is pale pink and contains small, very lightly stained *specific granules*. There are also a few larger, lavender *azurophil granules*. Some of the larger granules are *lysosomes*, which function in intracellular digestion. They are membrane-bound vesicles that contain *acid hydrolases*, i.e., hydrolytic enzymes with acid pH optima. Neutrophils are the body's first defense against invading microorganisms. At a wound site, for example, they leave the bloodstream and engulf bacteria that may be present.

Eosinophils. Eosinophils are named for the tendency of their cytoplasmic granules to stain with acidic dyes. They comprise only 1–5% of the leukocyte population. Eosinophils are round cells, 10–14 μm in diameter, and have a dark purple nucleus that usually has two large lobes connected by a thin strand or a band of chromatin. The cytoplasm is distinctive, being filled with many large, reddish-orange *eosinophil granules*. The eosinophil granules are true lysosomes since they contain acid hydrolases within a limiting membrane. It appears that eosinophils selectively phagocytize foreign proteins that are complexed with antibodies.

Basophils. Basophils are named for the tendency of their cytoplasmic granules to stain with basic dyes. They are the least frequent cell type; basophils comprise only 0.5% of the leukocyte population and, hence, may be difficult to find. In fixed smears, they are about 10 μm in diameter. The nucleus, which stains purple, may be round, indented, band, or lobed. The cytoplasm contains numerous large, dark purple *basophil granules*, which serve to distinguish this cell type. The basophil granules are not lysosomes, for instead of acid hydrolases, they contain histamine and heparin. The function of the basophils is not well understood.

Lymphocytes. Lymphocytes are the second most abundant type of white blood cell, comprising 20–40% of the leukocyte population. They vary considerably in size; most lymphocytes are small, 7–10 μm in diameter, though some may be as large as 15 μm. The nucleus is round and large, often with an indentation. It stains dark purple, and there are usually clumps of chromatin present. The cytoplasm stains blue. In the small lymphocytes there is only a thin layer of cytoplasm surrounding the nucleus. The primary function of lymphocytes is the production of antibodies in the immune response.

Monocytes. Monocytes are large cells, 12–15 μm in diameter, and comprise 3–8% of the leukocyte population. Many are round or oval, and some have blunt *pseudopods*. The general appearance is that of a large lymphocyte, but the two cell types are distinguishable by their staining. The monocyte nucleus, which is usually round or kidney-shaped, stains lightly and does not have many clumps of chromatin. The monocyte cytoplasm is a dull gray-blue and contains small, lilac-stained granules. Monocytes remain in the bloodstream only 1–2 days; they function mainly in the body tissues where, as *macrophages*, they phagocytize a variety of foreign substances.

Platelets. In addition to all the foregoing cell types there are the platelets. They are small, irregularly shaped cell fragments that have broken away from larger cells in the

FIGURE 2.1
Human blood cells as they appear in a fixed smear. Er, erythrocytes; N(seg), segmented neutrophil; N(band), band neutrophil; Eo, eosinophil; B, basophil; L, lymphocyte; M, monocyte; P, platelets. (Courtesy of Abbott Laboratories, from *The Morphology of Human Blood Cells*, 5th edition, ©1985.)

bone marrow. Platelets range in size from 1–4 μm and stain blue or purple. These numerous cell fragments play a major role in the blood clotting process.

PERIODIC ACID–SCHIFF REACTION

The periodic acid–Schiff reaction was first introduced for histological preparations by J. F. A. McManus in 1946. Figure 2.2 shows the steps in the staining of a carbohydrate (glycogen). The first step in the PAS reaction is treatment of the fixed cells with periodic acid (HIO_4). The HIO_4 oxidizes the 1,2-glycol grouping of sugars to a dialdehyde. The preparation is then stained with *Schiff's reagent*, a colorless liquid prepared by treating *basic fuchsin* (a red dye) with sulfurous acid. Schiff's reagent reacts with aldehydes to form colored compounds (purplish-red) by a mechanism that is not completely understood. Any cell structures that stain purplish-red with the PAS reaction are said to be *PAS-positive*. Although the PAS reaction cytochemically identifies carbohydrates, it also stains some lipids other than glycolipids, specifically, unsaturated lipids and phospholipids. The PAS reaction does not stain ribonucleic acid (RNA) or deoxyribonucleic acid (DNA) as their sugars lack the 1,2-glycol grouping (see Fig. 3.2, p. 25, for the structures of RNA and DNA).

Counterstaining.   Since the PAS reaction does not stain nucleic acids, the nucleus of the cell shows little or no staining. Accordingly, after the PAS reaction, a nuclear *counterstain* is commonly used. You will counterstain with *Harris's hematoxylin*, a mixture of hematein and ammonium aluminum sulfate. The hematoxylin, which has an affinity for chromatin, stains the nucleus dark blue.

Diastase-Treated Control Slide.   In many cytochemical reactions, control slides are used to verify the specificity of the staining. For the PAS reaction, you will run a control slide treated with the enzyme *diastase*, which digests starch and glycogen. An absence of staining after diastase treatment is confirmation of the presence of glycogen in PAS-positive regions of the blood cells.

HIO₄

Schiff's reagent

Purplish-red-colored compounds

FIGURE 2.2
The steps in the periodic acid–Schiff (PAS) reaction, shown here staining a portion of a glycogen molecule.

Slide Preparation. The procedure selected for the PAS reaction in blood cells is a modification of the method developed by F. G. J. Hayhoe and coworkers in 1964 (see Miale, 1982). You will be provided with two blood smears that have previously been *fixed* in a 9:1 mixture of ethanol:formalin. In cytological preparations, cells must be fixed so that their general morphology and internal structures will be preserved. If the cells were not treated with *fixative*, the hydrolytic enzymes of the lysosomes would be released, and, eventually, much of the cell would be digested. An effective fixative must render cell components insoluble, lest they be washed out during subsequent treatment. It should also prevent subsequent swelling or shrinkage of the cell contents. Often, the fixative improves staining by enhancing the affinity of cell components for the dye molecules. Most fixatives contain an alcohol plus one or more of the following: acetic acid, formalin, chloroform.

For the various steps in the PAS reaction, the fixed blood smears will be transferred to or stored in Coplin jars. As the slides are being processed, the slide surface with the cell preparation must never be allowed to rest against another slide. You will work with two fixed blood smears, one for the PAS reaction and one for the diastase control. After the control slide is treated with diastase, both slides are treated with HIO_4 followed by Schiff's reagent. The staining jar should be kept in the dark since Schiff's reagent deteriorates rapidly in the light. After treatment with Schiff's reagent, the slides are rinsed with a *bisulfite bleaching solution*. This rinse assures that the red coloration is due to the cytochemical reaction and not due to the presence of any basic fuchsin that may have formed during the staining procedure. Subsequently, the slides are counterstained. Only half of each blood smear will be counterstained, so that there will be a portion of each smear stained only by the PAS reaction.

The final step in the procedure is the mounting of a coverslip. The coverslip protects the cell preparation and allows immersion oil to be removed easily. The choice of mounting medium is often dictated by the last solvent used in making the preparation. Sometimes it depends on the stain, as certain stains are leached out of the tissue by some mountants. A popular mountant is *Permount* (Fisher), a synthetic resin that is soluble in xylene. The mountant will not be dry for several days at room temperature or overnight in a warm incubator. The slides will be examined before the mountant has dried, so extreme care should be taken not to get the sticky mountant on the oil-immersion lens. When the preparation is completely dry, excess mountant can be removed from the edges of the coverslip with solvent.

PROCEDURES

PERIODIC ACID–SCHIFF REACTION

1. Label the slides (on the frosted end) in pencil with your initials and the staining procedure: one slide, labeled *PAS*, will be subjected to the PAS reaction; the second slide, labeled *D + PAS*, will be treated with diastase prior to the PAS reaction. The pencil markings will also serve to identify the slide surface with the cell preparation.
2. Place the slide labeled *D + PAS* in a 1% solution of diastase in phosphate buffer, pH 6.0, at 20° C, for 30 min. The enzyme treatment and all staining procedures are done in Coplin jars; use a zigzag arrangement of up to nine slides per jar. Always use forceps to place slides in or to remove slides from Coplin jars. During the enzyme treatment, the slide labeled *PAS* can simply be left on the lab bench.
3. Rinse the diastase-treated slides under running tap water for 2 min, in a Coplin jar. Then replace with distilled water and pour it off. Unless otherwise noted, slides do not have to be dried between steps in the PAS reaction.
4. Place both slides (*PAS* and *D + PAS*) in a 1% aqueous solution of periodic acid for 10 min.
5. Rinse the slides under running tap water for 1 min (in a Coplin jar) and then rinse with distilled water, as in step 3.
6. Place the slides in Schiff's reagent* for 10 min. The staining jar should be kept in the dark.
7. Remove the slides from the staining jar with forceps and transfer them to a clean Coplin jar. Rinse the slides under running distilled water for several seconds.
8. Replace the distilled water with three changes of freshly prepared bisulfite bleaching solution (180 ml distilled water, 10 ml 10% Na or K metabisulfite solution, 10 ml 1 N HCl), 2 min in each bisulfite rinse.
9. Rinse the slides under running tap water for 5 min and then in distilled water. Wipe the back of each slide with a Kimwipe and allow to air dry in a near vertical position.

*CAUTION: *Be careful with Schiff's reagent. Although it is a colorless liquid, it will stain skin, clothing, and other surfaces purplish-red. Always keep the staining jar on paper towels or Benchkote.*

10. Place the air-dried slides in the hematoxylin staining solution (counterstain) for 2 min. Each Coplin jar should have only enough of the staining solution (about 15 ml) so that approximately half of the blood smear will be immersed when the slide is placed in the jar, and no more than four slides should be placed in each Coplin jar at one time.

11. After staining, remove each slide from the Coplin jar and immediately rinse under running distilled water for several seconds. Wipe the back of each slide and allow to air dry in a near vertical position.

12. When the slides are completely dry, dip each in xylene (in a fume hood) and add a drop of Permount and a coverslip. Position the coverslip so that it will cover all of the smear. Lower the coverslip slowly to minimize air bubbles and then gently press out the excess liquid with a paper towel.

REFERENCES

Lillie, R. D. 1977. *H. J. Conn's Biological Stains*, 9th ed., pp. 26–28, 259–266, 489–502. Williams & Wilkins, Baltimore.

Miale, J. B. 1982. *Laboratory Medicine Hematology*, 6th ed. pp. 116–210, 658–670, 867–869, 872. C. V. Mosby, St. Louis.

Pearse, A. G. E. 1968. *Histochemistry, Theoretical and Applied*, Vol. I, 3rd ed., pp. 307–322, 447–451. Little, Brown, Boston.

OBSERVATIONS AND QUESTIONS

1. Examine the Wright–Giemsa–stained preparation with the low power and oil-immersion objectives. Do not use any colored filters. Why not?

_____

2. Under oil-immersion, identify a neutrophil, lymphocyte, monocyte, eosinophil, basophil (if encountered), erythrocytes, and platelets.

3. Using the high-dry and oil-immersion objectives without any colored filters, now scan the *PAS* slide, both the portion that was counterstained and the portion that was not counterstained. Do not let the oil-immersion objective touch the sticky mountant. Identify neutrophils, small and medium lymphocytes, eosinophils, and platelets. It may not be possible to distinguish large lymphocytes from monocytes. What region of the leukocytes has the Schiff's reagent stained?

_____

What region of the leukocytes has the hematoxylin stained?

_____

4. Which type of leukocyte has the highest concentration of PAS-positive material?

_____

Which type of leukocyte has the lowest concentration of PAS-positive material?

_____

Locate an eosinophil. Describe the staining response of the eosinophil granules.

_____

5. In a region of the smear that was not counterstained, identify the erythrocytes. Where is the erythrocyte stained most intensely?

_____

6. Identify the platelets and describe their staining response.

_____

7. Examine the diastase-treated control slide. How does the PAS staining in the neutrophils, lymphocytes, eosinophils, and platelets compare with that previously observed on the *PAS* slide?

_____

_____

_____

What do your observations reveal about the amount of glycogen in each type of leukocyte?

_____

_____

8. Examine the erythrocytes in a region of the smear that was not counterstained. How does the staining compare to that observed on the PAS slide without diastase treatment?

_____

What do your observations from both slides reveal about the chemical nature of the PAS-positive material in erythrocytes?

_____

_____

9. What is an advantage of having a counterstained preparation?

_____

_____

_____

What is an advantage of having a preparation that is _not_ counterstained?

_____

_____

_____

CELLULAR NUCLEIC ACIDS

INTRODUCTION

The nucleic acids are the primary informational macromolecules in the cell. There are two classes of nucleic acid, *deoxyribonucleic acid*, or *DNA*, and *ribonucleic acid*, or *RNA*. They specify the amino acid sequences of all proteins, including, of course, enzymes. Since the repertory of enzymes and structural proteins determines the cell's biochemical and physiological potential, it is ultimately the nucleic acids that are responsible for the phenotype.

Nearly all the DNA is contained within the nucleus. *Messenger RNA* (*mRNA*) is synthesized in the nucleus and carries the code contained in the DNA to the cytoplasm, specifically, to the *ribosomes*. The ribosomes are composed of protein and a different type of RNA, termed *ribosomal RNA* (*rRNA*). The bulk of the cell's RNA is rRNA. Furthermore, a cell that is active in protein synthesis has a high concentration of cytoplasmic RNA. Also involved in the process of protein synthesis is yet a third type of RNA, *transfer RNA* (*tRNA*), which transports the individual amino acids to the ribosomes and decodes the nucleic acid message contained in the mRNA.

In this project, you will identify cytochemically both DNA and RNA in human blood cells, using the *methyl green–pyronin method*. With this method, you will be able to localize the two classes of nucleic acids and estimate their relative concentrations in different cell types.

STRUCTURE OF RNA AND DNA

The structures of RNA and DNA are shown in Figures 3.1 and 3.2. Both RNA and DNA are polymers of *nucleotides* (Fig. 3.1a), each consisting of a 5-carbon sugar (*ribose* in RNA and *2'-deoxyribose* in DNA) covalently bonded to a phosphate at the 5' end and a nitrogenous base (Fig. 3.1b) at the 1' end. Also shown are the polynucleotide chains (Fig. 3.2 a, b) and the DNA *double helix* (Fig. 3.2c), in which the two chains are held together by hydrogen bonds between *complementary bases*, i.e., between *adenine* (*A*) and *thymine* (*T*) and between *guanine* (*G*) and *cytosine* (*C*). Each RNA molecule is composed of only one polynucleotide chain. It also differs from DNA in that the base *uracil* (Fig. 3.1b) replaces thymine. The single chain of RNA folds back on itself in places so that there often are both single-stranded and double-stranded regions in the same RNA molecule. The double-stranded regions are stabilized by hydrogen bonds between complementary bases, as in DNA.

METHYL GREEN–PYRONIN METHOD

The methyl green–pyronin method was introduced in the 1940s by Jean Brachet, whose cytochemical studies of the amount of RNA in animal tissues were among the earliest to suggest a link between RNA and protein synthesis. The staining procedure for the methyl green–pyronin reaction is rather straightforward: the fixed tissue, on

(a) *Nucleotides*

Ribonucleotide   2'-Deoxyribonucleotide

Purine bases

Adenine (A)   Guanine (G)

(b)

Pyrimidine bases

Thymine (T)   Cytosine (C)   Uracil (U)

FIGURE 3.1
The building blocks of nucleic acids. The repeating unit in RNA and DNA is the nucleotide (a), which consists of a nitrogenous base, a pentose sugar, and a phosphate, all covalently bonded. In RNA, the 5-carbon sugar is ribose, while in DNA it is 2'-deoxyribose. The base is a purine or pyrimidine (b); A, G, T, and C occur in DNA, while A, G, U, and C occur in RNA. A related structure is the *nucleoside*, which consists only of the sugar and base, lacking the phosphate.

slides, is immersed in a mixture of the two dyes, after which the intensely stained tissue is partially destained with a water rinse and an alcohol solution. Methyl green has a preferential affinity for DNA, which stains green, and pyronin for RNA, which stains rose-red. Because of the specificity of the staining and the relative ease with which the procedure is carried out, the methyl green–pyronin method remains a useful tool for demonstrating cellular nucleic acids, especially for estimating the RNA content of different tissues.

The precise mechanism for the staining specificity of the methyl green–pyronin reaction is not known. One model has been proposed by Scott (1967), who analyzed the molecular interactions of each dye with the two nucleic acids. He found that pyronin and certain other dyes with a planar structure interact most strongly with single-stranded nucleic acids that have more freely accessible bases, as occurs in RNA. Methyl green, on the other hand, has a nonplanar structure that interacts best with nucleic acids that have an intact double helical structure, as occurs in native DNA.

In addition to its staining specicity, a useful feature of the methyl green–pyronin reaction is its stoichiometry; the higher the concentrations of DNA and RNA, the greater the intensity of the methyl green staining and pyronin staining, respectively. With a technique known as *microspectrophotometry*, the intensity of the staining can be measured quantitatively, and the amount of each nucleic acid thereby determined. The pyronin reaction is commonly used for determining the amount of RNA in cells. While methyl green staining has been used for determining the DNA content of cells, it is generally the *Feulgen reaction* (discussed in Project 14) that is used for such microspectrophotometric analyses.

RNase-Treated Control Slide. The usual way to confirm the staining specificity for RNA is to run two parallel slides: an untreated slide stained with the methyl green–pyronin mixture and a control slide pretreated with *ribonuclease (RNase)*, an enzyme that catalyzes the degradation of RNA. The purpose of the RNase control is to

FIGURE 3.2
Structure of RNA and DNA. Each molecule of RNA (a) is a single polynucleotide chain. Each strand in DNA has a similar structure (b), but native DNA consists of two polynucleotide chains in the form of a double helix (c). The two chains are held together by hydrogen bonds (dotted lines) between A and T and between G and C.

RNA
(a)

A single strand of DNA
(b)

DNA double helix
(c)

3.4 nm

0.34 nm

confirm that the structures stained with pyronin on the untreated slide do so because they contain RNA. The use of an RNase-treated control is so routine that the procedure is sometimes referred to as the *ribonuclease, methyl green–pyronin method* (Pearse, 1968).

Slide Preparation. The staining protocol selected for this project was developed by E. B. Taft in 1951 (see Clark, 1973). You will be provided with two blood smears that have been fixed in *Carnoy's fluid*, a 6:3:1 mixture of ethanol:chloroform:acetic acid. This fixative preserves nucleic acids and keeps the white blood cells affixed to the slide during the procedure. (Red blood cells hemolyze and are not visible.) As in Project 2, all the steps are carried out in Coplin jars. As the slides are being processed, the slide surface with the cell preparation must never be allowed to rest against another slide.

After one of the fixed blood smears has been treated with RNase, both slides are placed in the methyl green–pyronin staining solution. After staining, it is necessary to remove excess stain, a process termed *differentiation*. Each slide is rinsed briefly in distilled water and then placed in a 3:1 mixture of *tert*-butanol:ethanol. The *tert*-butanol–ethanol solution also dehydrates the preparation so that the slides can be transferred directly to xylene. After a second xylene rinse, a coverslip is mounted (see Project 2, p. 19).

PROCEDURES

METHYL GREEN–PYRONIN METHOD

1. Label the slides in pencil with your initials and the staining procedure: one slide, labeled *MGP*, will simply be stained with the double-dye mixture; the other slide, labeled *RNase + MGP*, will be pretreated with ribonuclease. The pencil markings will also serve to identify the slide surface with the cell preparation.

2. Place the slide labeled *RNase + MGP* in a 0.1% aqueous solution of RNase, pH 6.5–7.0, at 37° C, for 30 min. The enzyme treatment and all staining procedures are done in Coplin jars; use a zigzag arrangement of up to nine slides per jar. Always use forceps to place slides in or to remove slides from Coplin jars. During the enzyme treatment, the slide labeled *MGP* can simply be left on the lab bench.

3. Rinse the RNase-treated slides under running tap water for 2 min, in a Coplin jar. Then replace with distilled water and pour it off. Slides do not have to be dried between steps in the methyl green–pyronin method.

4. Place both slides (*MGP* and *RNase + MGP*) in the methyl green–pyronin staining solution for 30 min.

5. Rinse each slide under running distilled water for 2 to 3 sec. Then quickly drain each slide, wipe the back of each with a Kimwipe, and place in the *tert*-butanol–ethanol solution for 2 min, agitating each slide at the outset.

6. Drain each slide, wipe the back of each, and place in xylene (in a fume hood) for 2 min, agitating each slide at the outset. Transfer to a second xylene rinse (in a fume hood) for 2 min.

7. Remove each slide from the second xylene rinse. Before all the xylene has evaporated from the slide surface, add a drop of Permount and a coverslip. Position the coverslip so that it will cover all the smear. Lower the coverslip slowly to minimize air bubbles and then gently press out the excess liquid with a paper towel.

REFERENCES

Clark, G. 1973. *Staining Procedures*, 3rd ed., p. 155. Williams & Wilkins, Baltimore.

Pearse, A. G. E. 1968. *Histochemistry, Theoretical and Applied*, Vol. I, 3rd ed., pp 268–272. Little, Brown, Boston.

Scott, J. E. 1967. On the mechanism of the methyl green–pyronin stain for nucleic acids. *Histochemie* 9:30–47.

OBSERVATIONS AND QUESTIONS

1. Using the high-dry and oil-immersion objectives without any colored filters, scan the slide that shows the methyl green–pyronin staining. Do not let the oil-immersion objective touch the sticky mountant. Identify the different types of leukocytes: neutrophils, small and medium lymphocytes, and eosinophils. It may not be possible to distinguish large lymphocytes from monocytes. If necessary, refer to Figure 2.1, page 17.
   Which type of leukocyte has the most intense pyronin staining?

   _____

   Which type of leukocyte has the least intense pyronin staining?

   _____

2. Examine the RNase-treated slide and describe the pyronin staining in the leukocytes.

   _____

   What do your observations indicate about the location of RNA?

   _____

3. On the basis of staining intensity on the *MGP* slide, which type of leukocyte has

   the highest concentration of RNA? _____
   From what you know about the functions of the various types of white blood cells, give a reasonable explanation for the observed high concentration of RNA in these cells.

   _____

   _____

   _____

   Which type of leukocyte has the lowest concentration of RNA?

   _____

4. Compare the color of the nuclei in lymphocytes on the *MGP* and RNase-treated slides. How do they differ?

   _____

   _____

   Explain the difference.

   _____

   _____

   _____

5. On the RNase-treated slide, compare the intensity of the methyl green staining in small lymphocytes and large lymphocytes (or monocytes). How do they differ?

_____

Explain the difference.

_____

_____

_____

6. Suggest a method that would demonstrate the staining specificity of methyl green.

_____

_____

SPECTROPHOTOMETRIC ANALYSIS OF DNA AND RNA

INTRODUCTION

Many methodologies have been developed for extracting, characterizing, and quantifying the various classes of DNA and RNA. Nucleic acids are often identified by their distinctive absorption spectra. An absorption spectrum is a curve showing how much light a substance absorbs at various wavelengths. All nucleic acids absorb intensely in the ultraviolet (UV) range, a result of the conjugated double bonds in the purines and pyrimidines (see Fig. 3.1b). In the first part of this project, you will obtain the absorption spectrum for samples of DNA and RNA while learning the technique of spectrophotometry. The greater part of the project involves extracting DNA and RNA from bovine liver and determining the amount of each nucleic acid in the tissue. The extracted nucleic acids, which are in a degraded form, are quantified by converting the deoxyribose and ribose to colored compounds and measuring the intensity of the colors spectrophotometrically.

SPECTROPHOTOMETRY

Spectrophotometry is a method used to measure how much radiant energy a substance absorbs at different wavelengths. Pigments and other colored materials absorb in the visible range (380–760 nm), although many substances absorb maximally at shorter wavelengths (in the UV) or at longer wavelengths (in the infrared, IR). From the absorption spectrum of an unknown substance, one can often identify it or, at least, place it in a class of compounds. Especially useful for identification is the *absorption maximum*, the wavelength at which the absorption

FIGURE 4.1
Optical system of a spectrophotometer.
(*Courtesy of Bausch & Lomb.*)

spectrum exhibits a peak. Spectrophotometry can also be used for the *quantitative* analysis of a known compound. From measurement of light absorption at an appropriate wavelength, the concentration of the compound can readily be determined. In pure form, the nucleic acids can be quantified by absorption measurements in the UV range.

Spectrophotometer Components.   The instrument used to measure the absorption of a sample at different wavelengths is the spectrophotometer. The optical system of a relatively simple spectrophotometer, such as the Bausch & Lomb *Spectronic 20*, is shown in Figure 4.1. For absorption measurements between 340 and 1000 nm (near UV, visible, and near IR), the light source is a *tungsten lamp*. In spectrophotometers capable of analyses in the UV range (200–360 nm), there is also a *deuterium lamp*. The light is focused on a *diffraction grating*, which splits the light into its component colors or wavelengths. Different wavelengths are selected by changing the position of the grating with the *cam*. The final beam that strikes the sample is relatively monochromatic; the spectral slitwidth is 20 nm in the simplest spectrophotometers and considerably narrower in more sophisticated instruments. Any light not absorbed by the sample will strike the *phototube*, which converts the transmitted light energy to an electric current. The electric signal is displayed on a meter that is calibrated both in terms of how much light the sample absorbs and how much light it transmits. In lieu of meters, some spectrophotometers have digital readouts.

Transmittance and Absorbance.   The fraction of incident light transmitted by a sample is termed the *transmittance* (*T*). Using symbols, $T = I/I_0$, where I_0 is the intensity of the light that strikes the sample and I is the intensity of the light after it has passed through the sample. Transmittance is usually expressed as a percentage:

$$\%T = \frac{I}{I_0} \times 100$$

The *absorbance (A)* or *optical density* of the sample is a logarithmic function of *T*:

$$A = \log \frac{1}{T} = \log \frac{I_0}{I}$$

At 100% transmittance, $A = \log 1 = 0$. If a sample transmits 50% of the incident light, $A = \log (1/0.5) = \log 2 = 0.30$. The spectrophotometer usually has two scales, one calibrated in %T from 100 to 0 and the other in *A* units from 0 to ∞, though the last calibrated division is 2.0. In plotting spectral data, it is usually the absorbance that is plotted against wavelength or concentration.

Use of the Spectrophotometer.   Prior to reading the absorbance of a sample, it is necessary to adjust the spectrophotometer for a reference "blank," which contains all the substances in the sample *except* the material being measured. Since the material being measured is dissolved in a solvent, it is necessary to compensate for the absorbance of the solvent itself. It is also necessary to compensate for the loss of light caused by reflections from the surface of the *cuvette*, the special glass (quartz for UV analyses) vial that holds the liquid sample. In most spectrophotometers there is a *single beam* of light and a sample holder that accommodates one cuvette. Prior to measuring the absorbance of the sample, the blank is inserted and the spectrophotometer adjusted so that the scale reads 100%T ($A = 0$). Any change in absorbance when the sample is placed in the light path must be due to the material being measured. If additional samples are to be read *at the same wavelength*, it is not necessary to adjust for the blank each time. When the wavelength is changed, however, the instrument must be readjusted for the blank, since the light source does not emit all wavelengths at the same intensity and the phototube is not equally

FIGURE 4.2
A typical standard curve, which is a plot of absorbance versus known concentrations of the pure substance.

sensitive to all wavelengths. The more sophisticated spectrophotometers have a *double beam*; both the sample and blank are accommodated in the sample holder, and the correction for the blank is made automatically at each wavelength. Some double-beam spectrophotometers can also scan a range of wavelengths automatically and make a permanent record of the absorption spectrum on a strip-chart recorder.

Beer–Lambert Law. For many compounds, an important relationship exists between absorbance and two sample parameters. The relationship, known as the *Beer–Lambert law*, states that absorbance is directly proportional to *solute concentration (c)* and to the *length of the light path (l)*. The equation for the relationship is:

$$A = \alpha c l$$

where the constant, α, is the *absorption coefficient*. In the biological literature, the concentration and α are usually expressed on a weight basis. In this project, *c* is expressed in micrograms per milliliter (μg/ml), *l* in centimeters, and α in centimeters squared per microgram ($cm^2/\mu g$). Since the light path (interior diameter of the cuvette) is the same for all solutions, a plot of absorbance versus concentration gives a straight line with slope α. Such a plot using *known* concentrations of the pure compound is termed a *standard curve* (Fig. 4.2). As can be seen in Figure 4.2, the standard curve can be used to determine the concentration of the same compound in a solution of *unknown* concentration. The absorbance of the unknown solution is measured and the concentration read from the graph (Fig. 4.2, dashed lines). If the Beer-Lambert equation is written as

$$c = \frac{A}{\alpha l}$$

it is clear that concentration can also be determined just from the absorbance of the unknown solution and the value of α. The absorption coefficients for biological molecules can be determined experimentally or can be found in the literature. The length of the light path, *l*, is usually 1 cm.

EXTRACTION OF DNA AND RNA

For the present extraction of DNA and RNA from bovine liver, you will use a quick and simple method developed by W. C. Schneider (1957). The procedure consists of tissue *homogenization*, i.e., cell disruption, followed by extractions with trichloroacetic acid (TCA) and ethanol. The first treatment is with *cold* TCA, which precipitates nucleic acids, proteins, and lipids. The precipitate is separated from the acid-soluble

FIGURE 4.3
Diphenylamine reaction for DNA.

components by centrifugation; the *pellet*, which contains the unpurified nucleic acids, is retained, and the overlying liquid, or *supernatant*, is discarded. The precipitate is further purified with ethanol, which dissolves out the lipids. The final step is treatment with *hot* TCA, which dissolves out and partially degrades the DNA and RNA. There is some depurination and some cleavage of phosphodiester bonds.

COLORIMETRIC ASSAYS

The assay procedures are based on color reactions with the pentose moieties in the extracted nucleic acids. The concentrations of the sugars are determined by *colorimetry*, a procedure in which the concentration of a compound is measured by the intensity of the color produced when specific test reagents are added. In such colorimetric reactions, the intensity of the color, which is measured spectrophotometrically, is directly proportional to concentration over a limited range. Since the Beer–Lambert law applies, a standard curve can be constructed and the concentration of an unknown determined. The colorimetric assays you will use for DNA and RNA are, respectively, the *diphenylamine* and *orcinol reactions*.

Diphenylamine Reaction.   The diphenylamine reaction, which is depicted in Figure 4.3, is specific for 2-deoxypentoses. The reagent contains acetic and sulfuric acids. When heated with DNA, the acids cleave some of the phosphodiester bonds and hydrolyze the glycosidic linkages between the sugars and purines. In addition, the 2-deoxyribose is converted to ω-*hydroxylevulinyl aldehyde*, which reacts with diphenylamine to produce blue-colored compounds. The intensity of the blue color, which is measured at 600 nm, is directly proportional to the concentration of the sugar. You will carry out the diphenylamine reaction on four DNA solutions of known concentration, measure the absorbance of each, and plot a standard curve. By measuring the absorbance of the extract, you can obtain the DNA concentration from the standard curve.

Orcinol Reaction.   The orcinol reaction is depicted in Figure 4.4. Here, the reagent contains HCl. When heated with RNA, the acid cleaves some of the phosphodiester bonds and hydrolyzes the glycosidic linkages between the sugars and purines. The hot acid also converts the ribose to *furfural*, which, in the presence of ferric ions, reacts

FIGURE 4.4
Orcinol reaction for RNA.

with orcinol to produce green-colored compounds. The orcinol reaction is not as specific as the diphenylamine reaction since *all* pentoses, including the deoxyribose of DNA, will react to some extent. The color contribution of DNA in the orcinol reaction is only about 10% the intensity of the same concentration of RNA. We can determine the color contribution that is due just to the RNA in the extract by subtracting the color contribution of the known concentration of DNA.

For generating the standard curve for the orcinol reaction, which is measured at 660 nm, you will use four RNA solutions of known concentration. The RNA in the standards will be from bovine liver since the RNA from each species and tissue has a characteristic purine:pyrimidine ratio. The standard curve for the orcinol reaction cannot be used directly to determine the RNA concentration in the extract since both RNA and DNA contribute to the absorbance of the extract. The absorbance due to DNA is easily estimated after you have determined the concentration of DNA in the extract; it is about 10% the absorbance of the same concentration of RNA on the standard curve. Once you have the absorbance due to DNA, you can subtract it from the absorbance reading for the extract. Using the remaining absorbance, you can determine the concentration of RNA from the standard curve for the orcinol reaction.

PROCEDURES

For all procedures in this project, you should work in pairs. The general procedures given here assume the use of single-beam spectrophotometers. For a double-beam spectrophotometer, follow the procedure in the instruction manual provided with the instrument.

UV SPECTROPHOTOMETRY

1. Turn on the power switch, select the correct lamp source (deuterium lamp), and allow the spectrophotometer to warm up for at least 5 min.
2. Add the DNA and RNA stock solutions to separate quartz cuvettes. For the reference blank, add the saline–sodium citrate solution to a third quartz cuvette; this is the solution in which the DNA and RNA are dissolved.

3. Set the wavelength at 200 nm.

4. If your spectrophotometer requires a *dark current* adjustment, proceed as follows. With the light path blocked and the cover of the sample holder closed, adjust the dark current control knob so that the needle on the scale reads 0%T.

5. Insert the reference blank in the sample holder, close the cover, and adjust the absorbance so that the needle on the scale reads 100%T ($A = 0$). When inserting any cuvette, always check that (a) the index line or other marking on the cuvette is aligned with the center marking on the sample holder, (b) the cuvette is at least half full and the liquid free of bubbles, and (c) the surface of the cuvette is free of dirt and fingerprints.

6. Replace the blank with the DNA sample, close the cover, and read the absorbance from the scale. Record all absorbance readings on Data Sheet 4.1. Continue taking absorbance readings for the DNA sample at 10-nm intervals until you have measured the absorbance at 300 nm. Remember to correct for the reference blank at each wavelength.

7. Now measure the absorbance of the RNA sample from 200–300 nm at 10-nm intervals. Remember to correct for the reference blank at each wavelength. Record the absorbance readings on Data Sheet 4.1.

NUCLEIC ACID EXTRACTION

Since enough material for the entire class is obtained from one liver sample, the homogenization should be done by one designated team. The extraction procedure is carried out by each team.

Homogenization.

1. Weigh out 30 g of frozen calf liver, cut into 1- to 2-cm cubes.
2. To a chilled Waring blender jar, add 120 ml ice-cold distilled water.
3. Turn on the blender to the highest speed and then add the pieces of almost defrosted liver, one at a time, a few seconds apart. After all the pieces have been added, continue homogenizing for 1 min.
4. Dispense 2.0 ml of the liver homogenate* to each team, adding each aliquot to a 15-ml centrifuge tube.

Extractions with TCA and Ethanol.

1. Add 5.0 ml of ice-cold 10% TCA† to the centrifuge tube. Mix well by drawing the mixture in and out of a Pasteur pipet (fitted with a rubber bulb).
2. Place the tube in a clinical centrifuge and spin at the highest speed (approximately 1300 g) for 2 min.
3. With the Pasteur pipet, remove and discard the supernatant. If there is a layer of solid material floating on top of the supernatant, save it by inserting the Pasteur pipet beneath this layer when the supernatant is aspirated.
4. Add another 5.0 ml of ice-cold 10% TCA to the pellet and resuspend with the Pasteur pipet. Centrifuge and aspirate as described in steps 2 and 3, leaving one or two drops of liquid above the pellet.
5. With the Pasteur pipet, disperse the pellet in the liquid above it. Then, using a graduated cylinder, add 10 ml of 95% ethanol (room temperature) to the dispersed pellet. The clumps that form should be dispersed as much as possible with the Pasteur pipet.

---

*CAUTION: *Do not pipet by mouth.*

†CAUTION: *TCA is a strong acid; do not pipet by mouth. If any spills on your skin or clothing, wash it off immediately with water. In addition, protective goggles should be worn.*

6. Centrifuge at 1300 g for 2 min. Decant and discard the supernatant.

7. Wash the pellet in another 10 ml of 95% ethanol, as in step 5, and then centrifuge and decant as in step 6.

8. Add 5.0 ml of 5% TCA (room temperature) to the centrifuge tube and disperse the pellet with the Pasteur pipet. Place the tube in a water bath at 90° C for 15 min, agitating the tube every few min.

9. Centrifuge at 1300 g for 2 min and then carefully decant the supernatant, which contains the degraded nucleic acids, into a test tube.

10. To make sure all the remaining solubilized nucleic acid is removed, wash the pellet with 5.0 ml of 5% TCA (room temperature), centrifuge at 1300 g for 2 min, and then combine the supernatant with the nucleic acid extract in the test tube. This combined *nucleic acid extract* will be used for the colorimetric assays.

DIPHENYLAMINE REACTION

1. Label six test tubes as shown in the following table and add the indicated volume of each solution.* Tubes 1–4 are the standards; tube 5 contains the extract with an unknown concentration of DNA. The DNA stock solutions have 5% TCA as the diluent.

Tube	[DNA] (μg/ml) in each standard	Volume of each DNA stock solution*	Nucleic acid extract*	5% TCA*
Blank	0	—	—	2.0 ml
1	100	2.0 ml (100 μg/ml)	—	—
2	200	2.0 ml (200 μg/ml)	—	—
3	300	2.0 ml (300 μg/ml)	—	—
4	400	2.0 ml (400 μg/ml)	—	—
5	—	—	2.0 ml	—

*CAUTION: *TCA is a strong acid; do not pipet by mouth.*

2. To each tube, add 4.0 ml of the diphenylamine reagent[†] dispensed from a buret. Cover each tube with Parafilm and invert three times to mix the contents.

3. Remove the Parafilm, place loose-fitting screw caps on the tubes, and then place the tubes in a beaker of vigorously boiling water (with boiling chips) for 10 min.

4. While the tubes are in the boiling-water bath, turn on the spectrophotometer; it should be allowed to warm up for at least 5 min.

5. After the tubes have been in the boiling-water bath for 10 min, transfer them to an ice-water bath to cool them quickly.

6. Carefully transfer the liquid in the six test tubes to six labeled cuvettes. The cuvettes should be labeled near the top edge with a permanent marking pen.

7. Set the wavelength on the spectrophotometer at 600 nm.

8. With no cuvette in the light path and the cover of the sample holder closed, adjust the *dark current* control knob (termed "Power Switch/Zero Control" on the Spectronic 20) so that the needle on the scale reads 0%T.

9. Insert the reference blank in the sample holder, close the cover, and adjust the absorbance (with the "100%T Control" on the Spectronic 20) so that the needle on the scale reads 100%T ($A = 0$). When inserting any cuvette, always check that (a) the index line or other marking on the cuvette is aligned with the center marking on the sample holder, (b) the cuvette is at least half full and the liquid free of bubbles, and (c) the surface of the cuvette is free of dirt and fingerprints.

[†]**CAUTION:** *The diphenylamine reagent contains concentrated acetic acid and sulfuric acid. If any spills on your skin or clothing, wash it off immediately with water. In addition, protective goggles should be worn.*

10. Read the absorbance of each standard (tubes 1–4) and of the extract (tube 5). Since the readings are taken at the same wavelength, the spectrophotometer need not be adjusted for the blank between readings. Enter the absorbance readings on Data Sheet 4.2.

ORCINOL REACTION

1. Label six test tubes as shown in the following table and add the indicated volume of each solution.* Tubes 1–4 are the standards; tube 5 contains the extract with an unknown concentration of RNA. The RNA stock solutions have 5% TCA as the diluent.

Tube	[RNA] (μg/ml) in each standard	Volume of each RNA stock solution*	Nucleic acid extract*	5% TCA*
Blank	0	—	—	3.0 ml
1	100	0.4 ml (100 μg/ml)	—	2.6 ml
2	200	0.4 ml (200 μg/ml)	—	2.6 ml
3	300	0.4 ml (300 μg/ml)	—	2.6 ml
4	400	0.4 ml (400 μg/ml)	—	2.6 ml
5	—	—	0.4 ml	2.6 ml

*CAUTION: *TCA is a strong acid; do not pipet by mouth.*

The RNA concentrations for tubes 1–4 are taken to be those listed in the table; the addition of TCA is a required dilution in the orcinol reaction, which is an extremely sensitive colorimetric reaction.

2. To each tube, add 3.0 ml of the orcinol reagent[†] dispensed from a buret. Cover each tube with Parafilm and invert three times to mix the contents.
3. Remove the Parafilm, place loose-fitting screw caps on the tubes, and then place the tubes in a beaker of vigorously boiling water (with boiling chips) for 20 min.
4. After the tubes have been in the boiling-water bath for 20 min, transfer them to an ice-water bath to cool them quickly.
5. Carefully transfer the liquid in the six test tubes to six labeled cuvettes.
6. Set the wavelength on the spectrophotometer (warmed up) at 660 nm, adjust for the blank, and read the absorbance of each standard (tubes 1–4) and of the extract (tube 5). Enter the absorbance readings on Data Sheet 4.2.

REFERENCES

Adams, R. L. P., Burdon, R. H., Campbell, A. M., Leader, D. P., and Smellie, R. M. S. 1981. *The Biochemistry of the Nucleic Acids*, 9th ed. Chapman and Hall, London.

Schneider, W. C. 1957. Determination of nucleic acids in tissues by pentose analysis. In *Methods in Enzymology*, Vol. 3, Colowick, S. P. and Kaplan, N. O., eds., pp. 680–684. Academic Press, New York.

Segel, I. H. 1976. *Biochemical Calculations*, 2nd ed., pp. 324–346. John Wiley, New York.

Slayter, E. M. 1970. *Optical Methods in Biology*, pp. 528–551. John Wiley, New York.

[†]**CAUTION:** *The orcinol reagent contains concentrated hydrochloric acid. If any spills on your skin or clothing, wash it off immediately with water. In addition, protective goggles should be worn.*

EXERCISES AND QUESTIONS

1. On Graphs 4.1 and 4.2, plot the UV absorption spectrum for DNA and for RNA, respectively. Plot absorbance (ordinate) versus wavelength (abscissa). Each curve should be carefully drawn with a French curve template.

2. At what wavelength is absorption a maximum for DNA? _____ for RNA?

   _____

3. On Graph 4.3, plot the standard curve for the diphenylamine reaction. Plot absorbance (ordinate) versus the known DNA concentrations (μg/ml, abscissa) in the standards (tubes 1–4). The reference blank should also be included as a point ($A = 0$, [DNA] $= 0$). Using a ruler, draw the best-fit line for the points.

4. From your standard curve and the absorbance reading of the extract, determine the concentration of DNA in the extract. Enter the value on Data Sheet 4.2.

5. On Graph 4.4, plot the standard curve for the orcinol reaction. Plot absorbance (ordinate) versus the known RNA concentrations (μg/ml, abscissa) in the standards (tubes 1–4). The reference blank should also be included as a point ($A = 0$, [RNA] $= 0$). Using a ruler, draw the best-fit line for the points.

6. Determine the color contribution of DNA to the orcinol reaction of the extract. DNA reacts one-tenth as much as the same concentration of RNA. From the standard curve for the orcinol reaction, find the absorbance value for an RNA concentration that is the same as the concentration of *DNA* in the extract. Take one-tenth of that absorbance as the absorbance component due to DNA and enter the values.

   Concentration of DNA in the extract = _____

   Absorbance of RNA at concentration above = _____
   Absorbance due to DNA = 0.1 (absorbance of RNA at concentration above)

   = _____

7. To obtain the approximate concentration of RNA from the absorbance reading of the extract, it is necessary to subtract the absorbance component due to DNA.

   absorbance of extract _____

   − absorbance due to DNA _____
   _____

   absorbance due to RNA _____

   On the standard curve for the orcinol reaction, look up the concentration of RNA for the absorbance due to the RNA. Enter the value on Data Sheet 4.2.

8. Calculate the ratio [RNA]/[DNA] in the extract using the values obtained from the standard curves. _____ What trend in this ratio would you expect to find for tissues more active in protein synthesis? Explain.

_____

_____

_____

_____

9. What is the concentration of DNA and RNA in bovine liver? _____ μg DNA/mg liver; _____ μg RNA/mg liver. Show the calculations in the space provided.

10. Assuming that the light path is 1 cm, what is the value of α for the diphenylamine reaction? _____ cm²/μg. Show calculations.

Data Sheet 4.1

UV absorbance readings for DNA and RNA dissolved in saline–sodium citrate.

Wavelength (nm)	DNA	RNA
200		
210		
220		
230		
240		
250		
260		
270		
280		
290		
300		

<u>DNA</u>

Blank	Absorbance	T (%)
1	.235	58%
2	.51	31%
3	1.3	8.5%
4	above 2	—
5	.445	36%

<u>RNA</u>

Blank	A	T
1	.089	81.9%
2	.195	64%
3	.34	46%
4	.425	38%
5	.21	61.9%

Data Sheet 4.2
Absorbance readings for the diphenylamine and orcinol reactions

	Diphenylamine reaction	
Tube	[DNA] (μg/ml)	A_{600}
1	100	
2	200	
3	300	
4	400	
5 (extract)		

	Orcinol reaction	
Tube	[RNA] (μg/ml)	A_{660}
1	100	
2	200	
3	300	
4	400	
5 (extract)		

Graph 4.1
UV absorption spectrum for DNA.

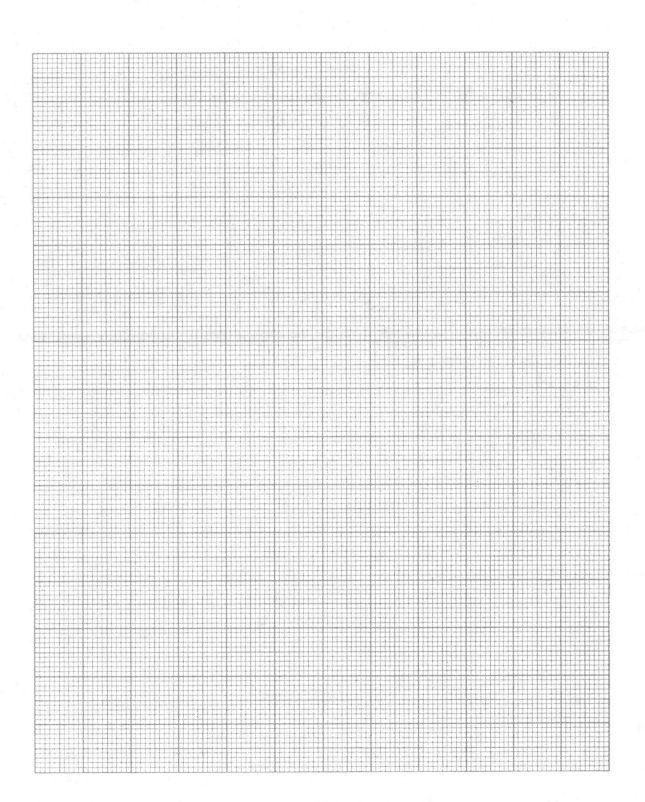

Graph 4.2
UV absorption spectrum for RNA.

Graph 4.3
Standard curve for the diphenylamine reaction.

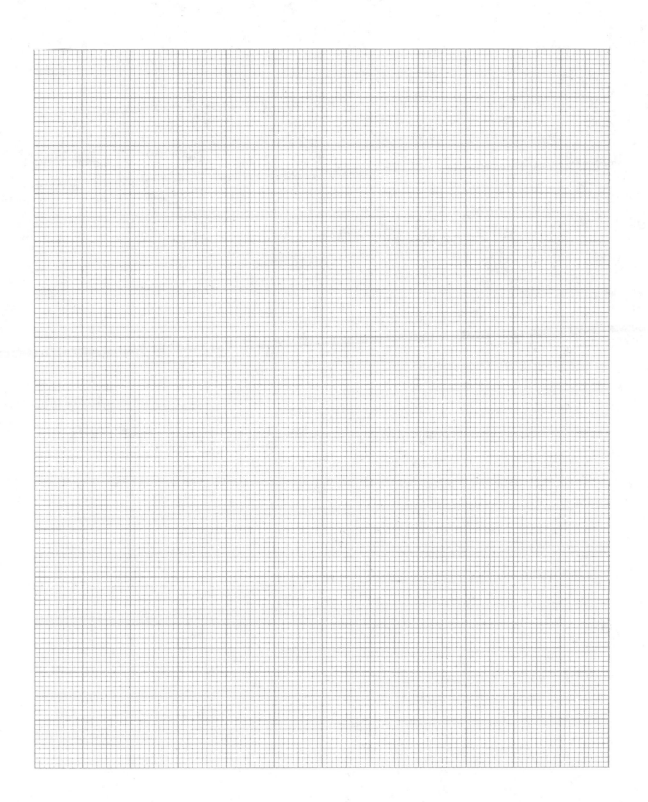

Graph 4.4
Standard curve for the orcinol reaction.

ELECTROPHORESIS OF HEMOGLOBIN

INTRODUCTION

Electrophoresis is a method commonly used for separating and characterizing charged molecules, especially proteins and nucleic acids. All electrophoretic methods are based on the principle that charged molecules will migrate through a liquid or semisolid medium that is subjected to an electric field. Each charged molecule migrates at a characteristic rate depending on its charge, size, and other physical characteristics. In the process, different charged molecules in a mixture migrate different distances and ultimately form discrete bands in the support medium. Electrophoresis is so discriminating a method that it can often separate two proteins that differ in only a single amino acid. One protein that has been studied extensively by electrophoresis is hemoglobin, the oxygen-transporting pigment in red blood cells. In this project, you will use *cellulose acetate electrophoresis* to separate normal and abnormal forms of human hemoglobin from supplied samples.

PROTEIN STRUCTURE

All proteins are polymers of amino acids. Figure 5.1 shows the general structure of amino acids and the peptide bond that links them together in proteins. Attached to the α-(alpha-) *carbon* are an α-*amino group*, an α-*carboxyl group*, a *hydrogen atom*, and a *radical* or *R group* (two different R groups, designated R_1 and R_2 are shown). The peptide bond forms between the α-carboxyl group of one amino acid and the α-amino group of another amino acid, resulting in a *dipeptide*. Chains of more than 10 amino acids are called *polypeptides*. There are 20 different amino acids commonly occurring in proteins. Each has a different R group, which not only distinguishes it from the

FIGURE 5.1
General structure of amino acids and formation of a dipeptide.

Valine

Nonpolar R group

Glutamic acid

Acidic R group

Lysine

Basic R group

$H_3N^+{-}CH_2{-}CH_2{-}CH_2{-}CH_2{-}\overset{\overset{\textstyle H}{|}}{\underset{\underset{\textstyle +}{\underset{\textstyle NH_3}{|}}}{C}}{-}COO^-$

FIGURE 5.2
Structures of valine, glutamic acid, and lysine.

other amino acids but also affects its chemical properties. The amino acids of particular interest in this project are depicted in Figure 5.2. They are *valine*, which has a nonpolar R group, *glutamic acid*, which has an acidic or negatively charged R group, and *lysine*, which has a basic or positively charged R group.

Some proteins are composed of one long polypeptide chain, while others contain two or more chains. The sequence of amino acids in the polypeptide is designated the *primary structure*. There are higher orders of structure that impart a three-dimensional shape, or *conformation*, to the protein. The *secondary structure* refers to the orientation of neighboring amino acids in three-dimensional space. A very common type of secondary structure is the α-helix, a right-handed coil that is stabilized by hydrogen bonds between neighboring amino acids. About 75% of the hemoglobin molecule is in an α-helix. In *globular proteins*, which include hemoglobin, other transport molecules, and enzymes, helical regions are separated by nonhelical

FIGURE 5.3
The quaternary structure of normal adult hemoglobin, which consists of four polypeptide chains, each associated with a heme group. There are two identical α chains and two identical β chains. Also depicted is a stretch of alpha helix in the β₁ chain. The letters N and C denote, respectively, the α-amino end and α-carboxyl end of each chain.

ELECTROPHORESIS OF HEMOGLOBIN

stretches of the polypeptide that twist and turn without any repeating structure. Such *random coils*, as they are termed, permit the helical regions to fold and twist, ultimately generating the *tertiary structure*, which is the conformation of the *entire* polypeptide. It is the sequence of amino acids in the polypeptide chain that determines the final conformation and function of the protein. Polar and charged amino acid residues, which are hydrophilic, remain at the surface of the molecule exposed to the aqueous environment, while nonpolar residues, which are hydrophobic, tend to cluster in the interior of the molecule. The tertiary structure is stabilized by electrostatic bonds between positively and negatively charged R groups, disulfide bridges, hydrogen bonds, and hydrophobic bonds between nonpolar residues. In the case of proteins composed of more than one polypeptide chain, the term *quaternary structure* is used to describe the orientation of the chains and the nature of the bonds that stabilize this orientation. Figure 5.3 shows the quaternary structure of normal adult hemoglobin, which contains four polypeptide chains.

HUMAN HEMOGLOBINS

There are five types of human hemoglobin that you will be separating in this project: hemoglobins A, A_2, F, S, and C.

Hemoglobin A.   Normal adult hemoglobin or *hemoglobin A* (*Hb A*) makes up 97–98% of the hemoglobin in the adult. Hb A contains two identical α-*globin* chains, each with 141 amino acids, and two identical β-*globin* chains, each with 146 residues. Thus, the formula for the Hb A tetramer can be written $\alpha_2\beta_2$. Each globin chain is associated with an iron-containing *heme group* (Fig. 5.3), a nonprotein component that is the site of oxygen binding.

Hemoglobin A_2.   *Hemoglobin A_2* (*Hb A_2*), a minor class (2–3%) of hemoglobin in the normal adult, contains two α chains and two δ (delta) chains, the latter having β-like sequences. The δ chain differs from the β chain in only 10 of its 146 amino acids. The formula for Hb A_2 is $\alpha_2\delta_2$.

Hemoglobin F.   The predominant hemoglobin from the eighth week of development until birth is *fetal hemoglobin* (*Hb F*). Hb F contains two α chains and two γ (gamma) chains, the latter being analogous to the β chains. The γ chain differs from the β chain in 39 of its 146 amino acids. Fetal hemoglobin can be represented as $\alpha_2\gamma_2$. By the time a normal infant is about six months old, very little Hb F remains in the blood. Only in individuals with certain inherited blood disorders in which the β chain is not synthesized is Hb F still produced in appreciable quantities throughout adult life.

Hemoglobin S.   The first abnormal hemoglobin to be studied at the molecular level was *sickle-cell hemoglobin* (*Hb S*), which is responsible for sickle-cell anemia, a serious heritable disease that occurs primarily in individuals of African ancestry. In afflicted individuals, the red blood cells have an abnormal morphology; instead of being biconcave disks, the erythrocytes are crescent- or sickle-shaped. The misshapen cells block small blood vessels, damaging internal organs, and are short-lived, resulting in anemia. At the protein level, the difference between Hb A and Hb S is but a single amino acid in each β chain. Specifically, at the sixth position from the α-amino end, glutamic acid is replaced by valine. If we designate the abnormal β chain as β^s, the Hb S tetramer can be written as $\alpha_2\beta_2^s$.

Hemoglobin C.   A much rarer variant is *hemoglobin C* (*Hb C*), which is responsible for hemoglobin C disease, a less serious condition than sickle-cell anemia. The molecular basis for Hb C also involves a change at the sixth position of each β chain; in lieu of a glutamic acid, there is a lysine. The Hb C tetramer can be written as $\alpha_2\beta_2^c$.

Table 5.1 gives the tetrameric formulas of the aforementioned hemoglobins and summarizes differences in the primary structure of the β chains and differences in *isoelectric points* (discussed shortly).

Table 5.1
Structural differences and isoelectric points of hemoglobins A, A_2, F, S, and C.

Hemoglobin	Chain formula	Sixth amino acid in β chain	pI
Hb A	$\alpha_2\beta_2$	Glutamic acid	6.93
Hb A_2	$\alpha_2\delta_2$	—	7.22
Hb F	$\alpha_2\gamma_2$	—	7.00
Hb S	$\alpha_2\beta_2^s$	Valine	7.10
Hb C	$\alpha_2\beta_2^c$	Lysine	7.28

The pI values are from J. A. Koepke et al., 1975, *Clin. Chem. 21*:1953. (Reprinted with permission from Clinical Chemistry.)

ELECTROPHORESIS

Electrophoretic separations are possible because charged particles in a solution will migrate when subjected to an electric field. Negatively charged molecules migrate toward the positive pole, or *anode*, while positively charged molecules advance toward the negative pole, or *cathode*. In *zone electrophoresis*, the charged particles migrate through a solid or semisolid, porous support medium. For the separation of hemoglobins, you will use cellulose acetate strips as the support medium.

The basic components of an electrophoresis apparatus are shown in Figure 5.4. The two major components are the *electrophoresis chamber*, which holds the support medium and the electrolyte solution (buffer), and the *power supply*. The power supply converts the AC (alternating current) of the line source to DC (direct current), the

FIGURE 5.4
Components of an electrophoresis system (end view) for separating hemoglobins and, above it, the electropherogram (face view).

54

latter required for electrophoresis, and allows a specific voltage to be selected. The support medium containing the sample is placed between two buffer reservoirs, which are connected to the electrodes of the power supply. With certain support media, including cellulose acetate, moistened wicks serve as electrical conductors from the buffer to the plate. Since the only electrical connection between the electrodes is the buffer in the pores of the support medium, the various charged molecules in the sample migrate to the anode or cathode. If each molecular species has a unique migration rate, at the end of the process there will be a series of bands, each containing a different type of molecule. The bands are visualized using stains or staining reactions based on the biological activity of the molecules in the bands. The final pattern of bands is termed an *electropherogram* (Fig. 5.4).

Mechanism of Separation.   Proteins are separated primarily due to differences in charge. The net charge depends on the number of acidic and basic amino acids in the protein *and* on the pH of the solution. Consider two proteins that differ in a single internal amino acid, say, a glutamic acid versus a lysine (see Fig. 5.2 for their structures). Since the α-amino and α-carboxyl groups of these amino acids are bonded to adjacent amino acids in the polypeptide, we need consider only the *extra* amino or carboxyl group in the radical. At a high pH (excess OH^- ions), the extra carboxyl group of glutamic acid would lose a proton and be negatively charged ($-COO^-$), whereas the extra amino group of lysine would be uncharged ($-NH_2$). While the two proteins in question may both have a net negative charge at a high pH, the one with a glutamic acid will be more negatively charged. If these two proteins differ in no other way, then during electrophoresis at a high pH, the protein with the glutamic acid will migrate to the anode more rapidly than the protein with the lysine.

For each protein there is some unique pH at which the net charge on the molecule is zero. If electrophoresis is carried out at this pH, termed the *isoelectric point* and symbolized by *pI*, the protein remains stationary. If the electrophoresis buffer has a pH above the pI of the protein, that protein will be negatively charged and migrate toward the anode, while if the pH is below the pI, it will be positively charged and migrate toward the cathode. The electrophoresis buffer used in this project has a higher pH (8.2–8.6) than any of the isoelectric points for the hemoglobins you will be studying (Table 5.1). Thus, the hemoglobin molecules will all be negatively charged and migrate toward the anode.

Let us examine the various factors that determine the rate at which a molecule migrates in an electric field. There is a *driving force* on the molecule. The driving force is the product of the net charge on the molecule and the *field strength*, defined as voltage/distance between the electrodes. Since the molecules are not migrating through a vacuum but through a buffer solution in the pores of the support medium, there will be a countering drag or *resisting force*. The resisting force is directly proportional to the size of the molecule, the viscosity of the buffer solution, and the velocity at which the molecule is moving. As electrophoresis proceeds, some final velocity is attained. It can be shown that the final velocity is directly proportional to the charge on the molecule and field strength but is inversely proportional to molecular size and viscosity of the medium.

Other factors that affect migration rate are the shape of the molecule and the nature of the support medium. As might be predicted, the less streamlined the molecule's shape, the slower will be its migration rate. Similarly, the greater the adsorption of the molecule to the matrix of the support medium, the slower will be its rate. Since there is very little adsorption to the cellulose acetate matrix, migration is not impeded, and separations can be completed in a relatively short time. Another important factor is *electroosmosis*, which refers to the tendency of the buffer to flow toward the cathode. As a result of this flow, the migration rate of negatively charged molecules is decreased, and the migration rate of positively charged molecules is increased. Molecules having no charge do not remain at the point of application but move toward the cathode, carried along by the buffer.

Electrophoretic Mobility.   A measurement occasionally used for comparing migration rates is *electrophoretic mobility*, μ (mu). Electrophoretic mobility is defined as the migration rate (cm/sec) of a particle at a field strength of 1 volt/cm. The formula used to calculate μ is

$$\mu = \frac{\text{migration rate in cm/sec}}{\text{field strength in volts/cm}}$$

Thus, the units for electrophoretic mobility are centimeters squared per second per volt ($cm^2 \, sec^{-1} \, volt^{-1}$). The sign of μ is that of the net charge on the particle. Protein molecules have relatively slow migration rates, with the absolute value of μ typically between 10^{-5} and $10^{-4} \, cm^2 \, sec^{-1} \, volt^{-1}$. You will be obtaining approximate values of μ for five different hemoglobins (A, A_2, F, S, and C) in three samples that contain different combinations of them. Your calculations of μ will be based on the distance traveled by each hemoglobin, the duration of the electrophoresis, the voltage applied, and the distance between the points where the cellulose acetate plate contacts the moistened wicks (the latter measurement taken to be the distance between the electrodes).

Cellulose Acetate Electrophoresis of Hemoglobin.   Cellulose acetate is a white, uniform, rather brittle material. It is supplied as plates, consisting of a thin (130 μm) cellulose acetate membrane bonded to a clear, flexible layer of *Mylar* (Du Pont), an inert plastic backing that makes the brittle cellulose acetate easier to handle. Cellulose acetate has a microporous structure much like that of a sponge, with pores several micrometers in diameter comprising about 80% of the volume. It is through these pores, which will be filled with buffer, that the compounds being separated migrate.

The electrophoresis chamber is prepared by adding buffer to the *outer* anodic and cathodic buffer reservoirs and by draping a buffer-soaked wick over the inner wall of each buffer reservoir (Fig. 5.4). Prior to applying the hemoglobin samples, the cellulose acetate plate is soaked in buffer to fill the pores of the cellulose acetate with buffer. Since each sample must be applied as a very narrow streak, you will use a specially designed applicator with tips made of very fine (and delicate!) wire. The applicator is loaded by depressing the applicator tips into sample wells containing the hemoglobin samples. The three samples used for this project are prepared mixtures of two or more different hemoglobins. *Sample 1* contains hemoglobins A and A_2; *sample 2* contains hemoglobins A, A_2, and S; and *sample 3* contains hemoglobins A, F, S, and C. Hemoglobins A_2 and C are not present in the same sample because they have approximately the same electrophoretic mobilities with the electrophoresis system you will be using. To apply the samples, the cellulose acetate plate is placed in an aligning base and the loaded applicator is set in position above it (Fig. 5.5). When the spring-loaded button on the applicator is pressed, the tips make contact with the

FIGURE 5.5
Application of samples to the cellulose acetate plate. See text for explanation.

plate, delivering their samples to the cellulose acetate. After a second application is superimposed on the first, the cellulose acetate plate is electrophoresed.

The plate is placed on the wicks, *cellulose acetate side down*, with the samples closer to the *cathodic* buffer reservoir. The chamber is covered and the samples electrophoresed at 350 volts for 25 minutes. *The high voltage used for electrophoresis demands that extreme caution be exercised; never touch the electrophoresis chamber or any of the electrical connections while the power is turned on.*

Following electrophoresis, the hemoglobin bands are stained with *ponceau S*, an anionic dye that, at low pH, reacts with positively charged amino groups of the protein. Background staining of the cellulose acetate is removed by soaking the plate in a dilute solution of acetic acid. Finally, the plate is dried in an oven. Before the plate is examined, it should be placed in a protective, clear plastic envelope to prevent chipping of the cellulose acetate.

In addition to the hemoglobin bands, there will also be a very faint band visible at the point of application and another faint band close to it for *carbonic anhydrase*, an enzyme in red blood cells that functions in the transport of carbon dioxide. The carbonic anhydrase is released from red blood cells during preparation of the hemoglobin sample.

PROCEDURES[1]

For all procedures, you will work in teams of four. Each team will electrophorese three samples: *sample 1* (A, A$_2$), *sample 2* (A, A$_2$, S), *sample 3* (A, F, S, C). Each team will have its own micropipetter, sample well plate, applicator, and cellulose acetate plate.

PREPARATION OF THE ELECTROPHORESIS CHAMBER

1. After making certain that the power supply is turned off and unplugged, add 100 ml of buffer to each outer reservoir (anodic and cathodic) of the electrophoresis chamber.
2. Moisten two wicks with buffer. Then, drape a wet wick over the inner wall of each buffer reservoir, making certain that each wick is immersed in the buffer.
3. Cover the chamber to prevent evaporation.

APPLICATION OF THE HEMOGLOBIN SAMPLES

Care should be taken when handling the hemoglobin samples since there is always the slight possibility that they are contaminated with infectious agents. Accordingly, disposable gloves should be worn, all work surfaces covered with Benchkote or paper towels, spills wiped up, the area disinfected, and all waste materials disposed of properly.

1. Familiarize yourself with the following equipment items: the Bufferizer, a set of two containers designed to immerse the cellulose acetate plate into buffer very slowly; the micropipetter and disposable capillary tubes; the sample well plate, the reservoir for the hemoglobin samples; the applicator; and the aligning base, the plastic holder that positions the cellulose acetate plate during sample application.
2. With a permanent marking pen, make a small identification mark at a corner of the cellulose acetate plate, on the Mylar (glossy) side.

[1]The electrophoresis procedure has been adapted, by permission, from a protocol developed by Helena Laboratories.

3. Place the plate in a rack and then place the rack in the lower chamber of the Bufferizer. Add buffer to the upper chamber and proceed to steps 4 and 5 while the plates are soaking. The plates should soak for about 5 min (after they are completely immersed in buffer) before the samples are applied.

4. Using a micropipetter, place 5.0 μl of each hemoglobin sample* (sample 1, sample 2, sample 3) in a separate well of the sample well plate. You will run the separations in duplicate, so that sample 1 will be added to wells 2 and 3, sample 2 added to wells 4 and 5, and sample 3 added to wells 6 and 7. The two end wells (1 and 8) should not be used since the samples on the edge of the plate tend to curve as their components separate. The easiest way to fill each well is to expel the drop at the tip of the capillary tube and then touch the drop to the well. Use a fresh capillary tube for each of the three samples and be sure to record which sample is in which numbered well. If the samples will not be applied within 2 min, cover the sample wells with a clean glass slide to prevent evaporation.

5. Prime the applicator tips as follows. Place the applicator in the brackets of the sample well plate and depress the tips three or four times in the hemoglobin samples. Remove the applicator from the brackets and touch the tips to a blotter. Replace the applicator in the brackets of the sample well plate and proceed quickly through the next three steps.

6. Remove the cellulose acetate plate from the soaking buffer and place it on a blotter, *cellulose acetate side up*. With another blotter, blot the cellulose acetate once firmly.

7. Place a drop of water on the aligning base and then place the plate, *cellulose acetate side up*, in the aligning base so that the top edge of the plate is on the line labeled CATHODE APPLICATION (Fig. 5.5). In this position, the samples will be applied closer to one end (cathodic end) of the plate. One person should apply the hemoglobin samples for each team.

8. Apply the hemoglobin samples as follows, keeping in mind that once the applicator tips are loaded, the samples must be applied within 15 sec. Load the applicator tips by depressing them into the hemoglobin samples three or four times. Immediately, place the applicator in the brackets of the aligning base, which contains the cellulose acetate plate, depress the tips, and hold for 5 sec. Reload the applicator tips and make a second application of the hemoglobin samples directly on top of the first. Proceed *immediately* to electrophoresis.

ELECTROPHORESIS

1. Remove the cover from the electrophoresis chamber.

2. Place the plate, *cellulose acetate side down*, on the wicks, with the samples closer to the *cathodic* ($-$) buffer reservoir. Set a clean glass slide on the plate (as a weight) to ensure solid contact with the wicks. Up to three plates can be electrophoresed simultaneously in the same chamber.

3. Cover the chamber and then plug in and turn on the power supply.[†] Electrophorese at 350 volts for 25 min.

STAINING

All staining materials should be kept on Benchkote or paper towels.

1. When electrophoresis is completed and the power supply switched off, unplug the power supply and remove the cover from the electrophoresis chamber.

*CAUTION: *The hemoglobin samples contain a small quantity of cyanide.*

[†]CAUTION: *While the power supply is turned on, do not touch the electrophoresis chamber or any of the electrical connections.*

2. Remove the cellulose acetate plate from the chamber and place it in a staining rack.
3. Place the rack in a staining dish containing ponceau S stain for 5 min.
4. Transfer the rack through three destaining baths of 5% acetic acid, 2 min in each bath, or until the background is white or nearly so. The rack should be agitated occasionally in each bath.
5. Dry the plate, *cellulose acetate side up*, in a 56°C oven for 10 min or until dry. Place the dried plate in a protective, clear plastic envelope since cellulose acetate chips rather easily.

REFERENCES

Bunn, H. F., Forget, B. G., and Ranney, H. M. 1977. *Human Hemoglobins*. W. B. Saunders, Philadelphia.

Gaál, Ö., Medgyesi, G. A., and Vereczkey, L. 1980. *Electrophoresis in the Separation of Biological Macromolecules*, pp. 15–18, 35–36, 41–45, 300–307. John Wiley, New York.

Sheeler, P. and Bianchi, D. E. 1987. *Cell and Molecular Biology*, 3rd ed., pp. 110–118. John Wiley, New York.

Whitaker, J. R., 1967. *Paper Chromatography and Electrophoresis*, Vol. I, pp. 1–49, 102–194. Academic Press, New York.

EXERCISES AND QUESTIONS

1. Identify the band for Hb A and the band for Hb A_2 in sample 1 (Hb A is present in greater quantity than Hb A_2). Also identify the point of application and the band for carbonic anhydrase.
2. Identify the band for Hb S in sample 2.
3. Identify the bands for Hb F and Hb C in sample 3.
4. Calculate electrophoretic mobility (μ) for all five hemoglobins. For the distance between the electrodes, use the distance between the points where the plate contacted the wicks. Assume that the voltage across this distance was 350 volts. To obtain the distance traveled by each hemoglobin, measure the distance from the point of application to the center of each band. Remember to include the sign of μ. Enter the values in the spaces provided and show your calculations.

Distance between electrodes = _____ cm

Time = _____ sec

Hemoglobin	Distance traveled (cm)	μ (cm^2 sec^{-1} volt^{-1})
A		
A_2		
F		
S		
C		

5. From the known amino acid differences in the β chains of hemoglobins A, S, and C (Table 5.1), give an explanation for their relative electrophoretic mobilities.

_____

_____

_____

_____

6. Are the relative electrophoretic mobilities of hemoglobins A, F, S, and C consistent with their pI's (Table 5.1)? _____ Explain.

_____

_____

_____

_____

7. Give a possible explanation for the fact that Hb A_2 and Hb C have different pI's (Table 5.1) yet have nearly identical electrophoretic mobilities.

_____

_____

_____

8. From the known amino acid differences in hemoglobins A, S, and C, and from your knowledge of the different levels of protein structure, speculate why Hb C results in a less severe condition than does Hb S.

_____

_____

_____

_____

_____

ELECTROPHORESIS OF LDH ISOZYMES

INTRODUCTION

Within the cells and tissues of the same organism, certain enzymes occur in multiple molecular forms. Such multiple forms of an enzyme are termed *isoenzymes* or *isozymes*. One of the best studied families of isozymes is *lactate dehydrogenase* (*LDH*), which exists in five common forms. LDH catalyzes the final reaction of anaerobic glycolysis: the reduction of pyruvate to lactate and the concomitant oxidation of NADH to NAD^+ (*nicotinamide adenine dinucleotide*). Each of the five LDH isozymes catalyzes the same reaction, but each has unique kinetic properties. The distribution of the five LDH isozymes in a particular tissue plays a significant role in its metabolism by determining the rate at which pyruvate is converted to lactate. In this project, you will obtain homogenates from several tissues of a rat and, from these homogenates, separate the LDH isozymes using the technique of electrophoresis. The electrophoresis will be carried out on an *agarose gel*. The LDH bands formed in the gel are made visible by a specific staining reaction that will enable you to determine the relative amounts of the five isozymes in each tissue.

THE ISOZYMES OF LACTATE DEHYDROGENASE

Each of the five major LDH isozymes consists of four polypeptide chains. Biochemical studies have revealed that each polypeptide can be of two types, designated the *M* and *H* chains. These designations stem from the predominance of the M chain in the LDH isozymes of skeletal muscle and the predominance of the H chain in the LDH isozymes of the heart. Since there are four chains in each isozyme molecule and two possible subunits (M or H), there are five possible combinations. When all four chains are of the H type, the resultant tetramer has the formula H_4, also referred to as *LDH-1*. The LDH isozyme with one M chain and three H chains has the formula MH_3 and is designated *LDH-2*. Similarly, the LDH isozymes with the other combinations of the M and H chains have the formulas M_2H_2 (*LDH-3*), M_3H (*LDH-4*), and M_4 (*LDH-5*).

Each LDH isozyme has a molecular weight of about 140,000 daltons, and each polypeptide chain has a molecular weight of about 35,000 daltons. Despite their similar molecular weights, the M and H chains have very different amino acid compositions and confer different kinetic properties on the LDH molecule. The rate of conversion of pyruvate to lactate increases with the number of M chains. Thus, M_4 (LDH-5) has the highest rate for converting pyruvate to lactate, while H_4 (LDH-1) has the lowest rate and is even inhibited by excess pyruvate. The other LDH isozymes have rates that are proportional to the number of M chains.

The relative amounts of the five LDH isozymes in a tissue correlate well with the metabolic requirements of the tissue. In skeletal muscle, where oxygen deprivation is a common occurrence, LDH-5 is the most abundant isozyme. Since LDH-5 can rapidly convert pyruvate to lactate, the predominance of this isozyme allows glycolysis and ATP synthesis to continue in the absence of oxygen. In cardiac muscle, which is a more aerobic tissue, LDH-1 and LDH-2 predominate. Since these isozymes convert pyruvate to lactate very slowly, the pyruvate is preferentially oxidized *aerobically* to CO_2 and H_2O.

Different profiles of the five LDH isozymes are also observed at different times during embryonic development. In the rat and mouse, for example, most early embryonic tissues, including heart muscle, have LDH-5 and LDH-4 as the predominant isozymes. As development proceeds there is a gradual shift in many tissues toward the isozymes with more H chains.

The relative amounts of the five LDH isozymes at a specific time during development or in a particular tissue are determined primarily by the proportions of the M and H subunits in each cell. It has been shown that if the single M and H chains are mixed in specific proportions *in vitro*, they aggregate to form the different LDH isozymes in proportions predicted by the laws of probability. Like all proteins, the synthesis of the M and H chains is directed by genes. In cells where there is a preponderance of one of the two chains (M or H), we have an example of differential gene activity, with the genes that encode the M and H chains being transcribed at different rates. There is also evidence that during cell differentiation, different LDH isozymes are selectively degraded in certain tissues.

Assays of LDH isozymes are important not only in studies of metabolic regulation and cell differentiation but in clinical evaluations, as well. For example, normally LDH-2 is the predominant LDH isozyme in human serum, with LDH-1 being the second most abundant form. Following a heart attack, however, the damaged heart tissue releases excess LDH-1, so that LDH-1 becomes the predominant LDH isozyme in the blood. Such enzyme assays are used not only for evaluating cardiovascular disease but as an aid in the diagnosis of blood and liver disorders.

PREPARATION OF THE TISSUE EXTRACTS

You will be separating the LDH isozymes from the heart, liver, and gastrocnemius (calf) muscle of the rat. The cell fractions can be prepared in the laboratory by thoroughly mincing each tissue and then homogenizing with cold distilled water in a motor-driven, *Potter–Elvehjem tissue grinder* (Fig. 6.1), which is a glass tube with a fitted Teflon (Du Pont) or glass pestle. As the rotating pestle moves toward the bottom of the tube, the tissue is forced upward into the narrow space between the pestle and the wall of the tube. The shearing forces that are generated in this process disrupt the cells without damaging organelles. Care must be taken that the tissue does not get too warm during the homogenization since heating could destroy enzymatic activity, particularly of LDH-5. Accordingly, during homogenization, the tube should be immersed periodically in an ice-water bath. The resultant homogenate is centrifuged

FIGURE 6.1
Potter–Elvehjem tissue grinder for homogenizing animal tissue. The shaft of the pestle is connected to the shaft of a motor.

Pestle

Glass tube

Suspension

64

to sediment tissue debris and nuclei. The supernatant (suspended particles and liquid above the nuclear pellet) contains the LDH isozymes along with many other cellular molecules and all organelles except the nucleus. Before it is applied to the support medium for electrophoresis, the supernatant must be diluted. The diluted supernatant will hereafter be referred to as the *tissue extract*.

AGAROSE GEL ELECTROPHORESIS

The principles underlying the electrophoretic separation of the LDH isozymes are the same as discussed in Project 5, page 54, for hemoglobins. The differences in the two separations are strictly procedural. The electrophoresis chamber and power supply are the same as in Project 5, but the setup within the electrophoresis chamber is different because of the different support medium. For the separation of LDH isozymes in this project, the support medium is a thin layer of agarose gel affixed to a sheet of Mylar (Du Pont). The agarose plate yields an electropherogram that is easily handled and has a transparent background for *densitometry* (discussed shortly). Agarose is a polysaccharide that forms a porous gel, the diameter of the pores being a function of the agarose concentration. It is purer and much less acidic than ordinary agar, from which it is derived.

As with other types of electrophoresis, each sample must be applied to the support medium as a narrow streak. To apply the samples to the agarose gel, we use a cellophane template with narrow slits. With the template positioned on the agarose, several microliters of the diluted supernatant are placed in each slit (Fig. 6.2). After the samples have had time to diffuse into the agarose, the plate is ready for electrophoresis. The ends of the agarose gel plate will be immersed *directly* in the buffer (pH 8.1-8.3). Accordingly, the buffer is added to the *inner* buffer reservoirs of the electrophoresis chamber, and the agarose gel plate is flexed so as to fit in these wells. Thus, wicks are not required (as they were for the cellulose acetate plates in the separation of hemoglobins). The plate should be oriented with the *agarose side down* and the samples closer to the *cathodic* buffer reservoir. The chamber is covered and the samples electrophoresed at 100 volts for 15 minutes. *The high voltage used for electrophoresis demands that extreme caution be exercised; never touch the electrophoresis chamber or any of the electrical connections while the power is turned on.* Each of the five LDH isozymes has a different electrophoretic mobility, with the final sequence from anode (+) to cathode (−) being LDH-1, LDH-2, LDH-3, LDH-4, LDH-5.

VISUALIZATION AND QUANTIFICATION OF THE LDH BANDS

The tissue extract applied to the agarose gel contains not only the five LDH isozymes but many other proteins and cellular molecules, all colorless. To identify the LDH bands, we selectively stain them after electrophoresis. The agarose plate is incubated with a reaction mixture containing lithium lactate, NAD^+, nitro blue tetrazolium, and phenazine methosulfate. The lithium lactate serves as a substrate for the enzyme.

FIGURE 6.2
Application of samples to the agarose gel plate. See text for explanation.

Wherever there is LDH activity in the agarose gel, the lactate in the reagent is converted to pyruvate, and the NAD^+ is reduced to NADH, as summarized by

$$\text{lactate} + NAD^+ \xrightarrow{\text{LDH}} \text{pyruvate} + NADH + H^+$$

The hydrogens from the reduced NAD are transferred by phenazine methosulfate to nitro blue tetrazolium, reducing it to an insoluble, purple formazan dye, as follows:

$$NADH + H^+ + \text{nitro blue tetrazolium} \xrightarrow{\substack{\text{phenazine} \\ \text{methosulfate}}} NAD^+ + \underset{\text{(purple)}}{\text{formazan dye}}$$

Thus, each LDH band will be stained purple. To remove background staining, the agarose plate is soaked in a dilute solution of acetic acid. The agarose gel is then blotted with an acetic acid–moistened blotter and dried in an oven.

The foregoing staining reaction is stoichiometric; that is, the intensity of the purple color is proportional to the amount of LDH present. Any striking differences in staining intensity among the five bands can be readily detected by visual examination.

DENSITOMETRY

For accurate quantification of the LDH present in each band of the electropherogram, a *densitometer* is required. A densitometer is an instrument that uses either transmitted or reflected light to measure the optical density and, hence, the concentration of substances in bands on electropherograms and chromatograms. The electropherogram is scanned with a narrow beam of light at an appropriate wavelength (595 nm for the purple-stained LDH bands). As the transmitted light strikes a stained region (band) in the otherwise transparent agarose gel, the densitometer measures photoelectrically and records graphically how much light is absorbed by the band. In the final graph (Fig. 6.3), there is a series of steep curves, or peaks, one for each band. The area under each of the five peaks is proportional to the amount of LDH in the corresponding band. Measurement of the areas under the five peaks in Figure 6.3 indicates that the relative amounts of LDH-1, LDH-2, LDH-3, LDH-4, and LDH-5 are 5.2%, 5.7%, 4.7%, 1.6%, and 82.8%, respectively, in a commercially prepared extract of rat liver.

PROCEDURES[1]

PREPARATION OF THE TISSUE EXTRACTS

For the entire project, you should work in teams of four. Each team will be responsible for obtaining one of the tissue extracts for the entire class. Every team will then be able to electrophorese extracts from four different tissues: heart, liver, gastrocnemius muscle, and one other tissue. For all tissue homogenization procedures, the solutions and containers must be ice-cold to maintain the integrity of the lactate dehydrogenase.

    1. Sacrifice an adult rat with a sharp blow to the head or by decapitation.

    2. Wearing disposable gloves, dissect out the organ that has been assigned to your

---

[1]The electrophoresis procedure has been adapted, by permission, from a protocol developed by Helena Laboratories.

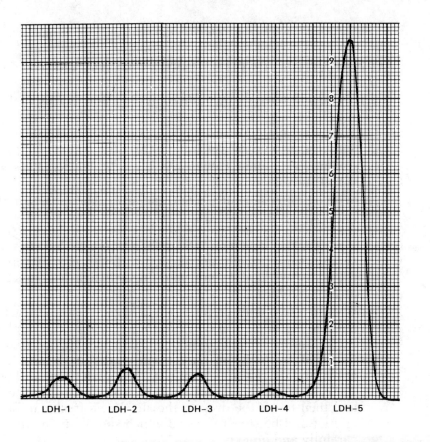

FIGURE 6.3
Densitometric tracing of an electropherogram of the LDH isozymes in rat liver.

LDH-1   LDH-2   LDH-3   LDH-4   LDH-5

team: heart, liver, gastrocnemius muscle, spleen, kidney, or lung. For all procedures, each tissue is to be kept separate from the others.

3. Rinse the tissue with ice-cold distilled water to remove as much blood as possible. Drain and weigh the tissue, recording the weight. Use no more than 2 g of tissue.

4. Place the tissue on a wooden chopping block and mince the tissue finely with a single-edge razor blade.

5. Place the minced tissue in the tube of a motor-driven, Potter–Elvehjem tissue grinder and add 1.0 ml of ice-cold distilled water for each gram of tissue.

6. Start the motor and then, with the rotating pestle in the tube (Fig. 6.1), move the tube up and down. To prevent the tissue from getting too warm during the homogenization, periodically place the tube in an ice-water bath. Continue the process until the tissue is homogenized thoroughly, or as much as seems possible.

7. Transfer the homogenate to a centrifuge tube and centrifuge at 1300 g for 10 min at 0°–4° C. Make sure that the centrifuge tubes are balanced; the tubes opposite each other should have the same total volume.

8. Transfer 1.0 ml of the supernatant* to an Erlenmyer flask and dilute with 49 ml of ice-cold distilled water. Mix well and then dispense the diluted supernatant (tissue extract) in a small vial to each team.

PREPARATION OF THE ELECTROPHORESIS CHAMBER

1. After making certain that the power supply is turned off and unplugged, add 35 ml of buffer to each inner reservoir (anodic and cathodic) of the electrophoresis chamber.

*CAUTION: Do not pipet by mouth.

2. Cover the chamber to prevent evaporation.

APPLICATION OF THE TISSUE EXTRACTS

Each team will have its own micropipetter and agarose gel plate.

1. Place a gel plate,* *agarose side up*, on a blotter D. The numbers 1–8 (for the eight samples) should be readable across the top of the plate. With a permanent marking pen, make a small identification mark at the corner of the plate, on the Mylar (glossy) side.

2. With a blotter A, gently blot the agarose where the samples will be applied, i.e., between the $\bar{0}$ demarcations. Then discard blotter A.

3. Place a cellophane template on the agarose so that the template slits, which will serve as sample wells, are aligned with the zero (0) demarcations at the sides of the plate (see Fig. 6.2). Run your fingertip lightly across the template to press out any air bubbles.

4. Using a micropipetter, place 3.0 μl of each tissue extract in a separate template slit (see Fig. 6.2). You will run the separations in duplicate, so that the tissue extract from heart will be added to slits 1 and 2, the tissue extract from liver will be added to slits 3 and 4, the tissue extract from muscle will be added to slits 5 and 6, and the tissue extract from the fourth tissue will be added to slits 7 and 8. The easiest way to fill each slit is to expel the drop at the tip of the capillary tube and then touch the drop to the slit. Use a fresh capillary tube for each tissue extract and be sure to record which sample is in which numbered slit. Work carefully and quickly.

5. After the last sample has been dispensed, wait 3 min; then gently blot the template with a fresh blotter A.

6. Remove and discard blotter A, wait 30 sec, and then remove and discard the template and blotter D. Proceed *immediately* to electrophoresis.

ELECTROPHORESIS

1. Remove the cover from the electrophoresis chamber.
2. With the *agarose side down* and the samples closer to the *cathodic* (−) buffer reservoir, place the ends of the gel plate into the inner buffer reservoirs, curling the plate so that it will fit. Two plates can be electrophoresed simultaneously in the same chamber.
3. Cover the chamber, making certain that the cover does not touch the gel plate(s), and then plug in and turn on the power supply.[†] Electrophorese at 100 volts for 15 min.

VISUALIZATION OF THE LDH BANDS

1. Immediately after electrophoresis is begun, prepare an incubation chamber (plastic case) as follows. Place a paper towel moistened with water in the incubation chamber and then place the closed chamber in a 45° C incubator until needed in step 7.
2. When electrophoresis is completed and the power supply switched off, unplug the power supply and then remove the cover from the electrophoresis chamber.

*CAUTION: *The agarose gel plate contains sodium azide, which is poisonous. When handling the plate, keep hands away from mouth. After handling the plate, wash your hands.*

[†]CAUTION: *While the power supply is turned on, do not touch the electrophoresis chamber or any of the electrical connections.*

3. Place a clean glass development slide (7.6 cm × 10 cm) on a paper towel.

4. Remove the gel plate from the electrophoresis chamber and place it, *agarose side up*, on the glass development slide. The anode edge of the plate should be closer to you.

5. Pour the entire vial of reconstituted LDH isozyme reagent along the cathode edge of the agarose.

6. Use a 5-ml pipet to spread the reagent across the agarose as follows. Lay the pipet lengthwise along the cathode edge of the plate. Spread the reagent by gently and slowly pulling the pipet toward you, being careful not to roll the reagent off the plate. Wait 15 sec, then reverse the direction of the spreading. Gently and slowly, move the pipet away from you, again being careful not to roll the reagent off the plate. Wait 15 sec and then spread the reagent toward you, this time allowing the reagent to roll off the plate.

7. Place the gel plate, *agarose side up*, in the preheated incubation chamber.

8. Place the closed incubation chamber in a 45° C incubator for 25 min.

9. After the incubation period, transfer the gel plate, *agarose side up*, to a large finger bowl containing 10% acetic acid, 1–2 cm deep, and gently swirl the solution for 2 min.

10. Transfer the gel plate, *agarose side up*, to a clean glass development slide. Then, stack the following on the plate: one blotter B moistened with 10% acetic acid (on the agarose), three folded paper towels (on blotter B), and, to provide pressure, a development weight (on the paper towels).

11. After 5 min, remove the weight, discard the blotter and paper towels, and then dry the gel plate by placing it, *agarose side up*, in a 60° C oven for 5 min or until dry. The electropherogram can now be evaluated visually or with a densitometer (at 595 nm), if one is available. For visual examination and storage, keep the dried plate in a clear plastic envelope.

REFERENCES

Everse, J. and Kaplan, N. O. 1973. Lactate dehydrogenases: structure and function. In *Advances in Enzymology*, Vol. 37, Meister, A., ed., pp. 61–133. John Wiley, New York.

Gaál, Ö., Medgyesi, G. A., and Vereczkey, L. 1980. *Electrophoresis in the Separation of Biological Macromolecules*, pp. 53–65, 178–186. John Wiley, New York.

Markert, C. L. and Ursprung, H. 1971. *Developmental Genetics*, pp. 36–49. Prentice-Hall, Englewood Cliffs, N.J.

Moss, D. W. 1982 *Isoenzymes*. Chapman and Hall, London.

NAME _____ SECTION _____ DATE _____

LAB PARTNERS _____

EXERCISES AND QUESTIONS

1. Examine the electropherogram and locate the bands containing LDH-1, LDH-2, LDH-3, LDH-4, and LDH-5.

2. Which LDH band(s) migrated toward the cathode? _____
   What are the possible causes for this movement?

   _____

   _____

   _____

3. For the heart, give the sequence of the five isozymes, from highest concentration to lowest concentration.

   _____   _____   _____   _____   _____
   Highest   Lowest

4. For the liver, give the sequence of the five isozymes, from highest concentration to lowest concentration.

   _____   _____   _____   _____   _____
   Highest   Lowest

5. For skeletal muscle, give the sequence of the five isozymes, from highest concentration to lowest concentration.

   _____   _____   _____   _____   _____
   Highest   Lowest

6. Name the fourth tissue and give the sequence of the five isozymes, from highest

   concentration to lowest concentration. _____

   _____   _____   _____   _____   _____
   Highest   Lowest

7. How do the LDH profiles in the heart and skeletal muscle differ?

   _____

   _____

   _____

   _____

   Which of these two profiles (heart or skeletal muscle) does the LDH profile from

   liver more closely resemble? _____
   Which of these two profiles (heart or skeletal muscle) does the LDH profile from

   the fourth tissue more closely resemble? _____

8. In which tissue(s) is there primarily only one of the two polypeptide chains (M or H) present? _____

Suggest events that might be occurring at the molecular level (transcriptional, posttranscriptional) to explain such LDH profiles.

_____

_____

_____

_____

9. If a densitometer is available, determine the percentage of each LDH isozyme in each tissue. Enter the values on Data Sheet 6.1.

Data Sheet 6.1
Percentage of each LDH isozyme in heart, liver, skeletal muscle, and the fourth tissue (identify in table).

Tissue	LDH-1	LDH-2	LDH-3	LDH-4	LDH-5
Heart					
Liver					
Muscle					

MEMBRANE PERMEABILITY

INTRODUCTION

Every cell is separated from its external environment by a *plasma membrane*. This thin (7.5 nm), primarily lipid–protein structure not only surrounds and contains the living protoplasm but regulates what substances enter and leave the cell. Investigations of the plasma membrane date back to the latter part of the nineteenth century when physiologists discovered its *selective permeability*. They observed that different substances penetrate the cell at different rates and that some substances do not penetrate at all. In 1899, E. Overton demonstrated that the greater the lipid solubility of a compound, the greater is its rate of penetration. This correlation, sometimes referred to as *Overton's rule*, provided one of the earliest indications that lipids are a major component of the plasma membrane. In this project, you will test alcohols with different lipid solubilities for their relative abilities to penetrate *Elodea* leaf cells.

MEMBRANE STRUCTURE

Today, the most widely accepted model for membrane structure is the *fluid mosaic model* (Fig. 7.1), proposed by S. J. Singer and G. L. Nicolson in 1972. Here, as in previous hypotheses, there is a *lipid bilayer*, with the hydrophobic, nonpolar "tails" of the phospholipid molecules pointing inward and the hydrophilic, polar "heads" of the phospholipids facing outward. What makes the fluid mosaic model unique is the dynamic relationship between the lipid bilayer and membrane proteins. The proteins are not in static layers, as previous models had proposed, but are completely interspersed with the lipids, producing a mosaic. The model portrays the lipid bilayer as a rather fluid structure in which the various proteins can move laterally. Two classes of proteins are proposed: the *integral proteins*, which penetrate the lipid bilayer partially or completely, and the *peripheral proteins*, which are attached to the surface. Some of the integral proteins are glycoproteins, having carbohydrate chains attached to them. Some of the lipids also have carbohydrate chains attached; these are the glycolipids. The carbohydrate chains always protrude from the cell surface to the exterior and, in certain cell types, confer antigenic specificity to the cell.

OSMOSIS IN PLANT AND ANIMAL CELLS

In many cells, such as *Elodea* leaf cells, the plasma membrane is freely permeable to water. The movement of water into or out of cells occurs by *diffusion*, a process of molecular movement from a region of higher concentration to a region of lower concentration due to the kinetic energy of the molecules. The diffusion of water across a membrane in response to a concentration gradient is termed *osmosis*. When a plant cell is placed in pure water, the water flows inward by osmosis. The cell volume

75

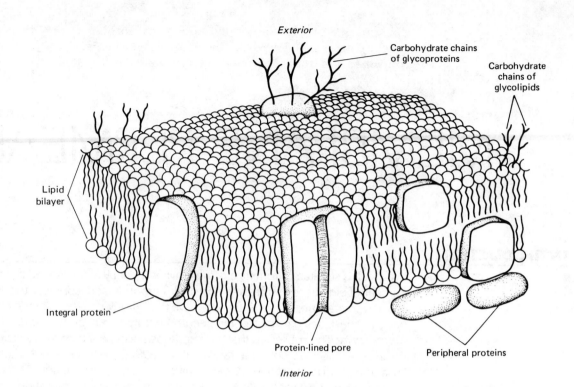

FIGURE 7.1
Fluid mosaic model of
membrane structure.

Labels in figure: Exterior; Carbohydrate chains of glycoproteins; Carbohydrate chains of glycolipids; Lipid bilayer; Integral protein; Protein-lined pore; Peripheral proteins; Interior

increases, causing the plasma membrane to expand against the relatively rigid cell wall. The resultant *osmotic pressure* increases until it prevents any further net flow inward, at which point an equilibrium is reached. When an animal cell such as an erythrocyte is placed in pure water, the consequences are different. As in the plant cell, water enters by osmosis, and the plasma membrane expands. Without a restraining cell wall, however, the membrane of the erythrocyte tears and the cell contents are released, a result known as *hemolysis*.

Both the cytoplasm and the liquid that bathes the cell are aqueous solutions; water is the *solvent*, and the proteins, sugars, and other dissolved material comprise the *solute*. The *solute concentration* is defined as the total number of dissolved particles per unit volume of solution. The terms *hypertonic, hypotonic,* and *isotonic* are used to describe relative solute concentrations. If the solute concentration outside the cell is *greater* than that inside, the external solution is described as *hypertonic*. Conversely, if the solute concentration outside the cell is *less* than that inside, the external solution is said to be *hypotonic*. A solution with the *same* solute concentration as the cell is *isotonic*.

Plasmolysis and Deplasmolysis in *Elodea* Leaf Cells.   Let us consider the situation of an *Elodea* leaf placed in an aqueous solution in which the solute penetrates the plasma membrane at a negligible rate. One such compound is sucrose. If the leaf is placed in an isotonic sucrose solution, there will be no net flow of water into or out of the cells because the concentration of the water (solvent) is equal inside and outside. If, however, the leaf is placed in a hypotonic sucrose solution, water will enter the cells by osmosis in response to the concentration gradient. The osmotic pressure will increase, though not as much as in pure water. If the leaf is placed in a hypertonic sucrose solution, water will leave the cells by osmosis. As a result of the net flow of water out of the cell, the plasma membrane and cell contents shrink away from the cell wall. This process is termed *plasmolysis*. (The plasma membrane is *not* lysed or disrupted.) If the extent and duration of plasmolysis are not too great, there may be no permanent damage. Recovery can occur if the plasmolyzed cell is transferred to a hypotonic solution; water flows into the cell, and *deplasmolysis* occurs. After recovery, water continues to enter by osmosis until equilibrium is reached. You will measure the tonicity of the *Elodea* cytoplasm by placing leaves in progressively higher concentrations of sucrose. The isotonic concentration lies between the highest concentration that prevents plasmolysis and the lowest concentration in which plasmolysis is observed.

PENETRATION OF NONELECTROLYTES

In this project, you will be measuring the rates at which various alcohols penetrate the plasma membrane of *Elodea* leaf cells. Nonelectrolytes generally penetrate by *passive transport*, where the driving force is the concentration gradient across the membrane. The *rate* of penetration is determined by several factors. Generally, the rate of penetration increases with increasing lipid solubility of the permeant. The lipid solubility of a compound can be specified by the ratio of its solubility in an organic solvent to its solubility in water. A frequently used measure of lipid solubility is the *ether:water partition coefficient*, defined as (solubility in diethyl ether)/(solubility in water). Table 7.1 gives the ether:water partition coefficients of the alcohols you will be testing.

TABLE 7.1
Ether:water partition coefficients (solubility in diethyl ether)/(solubility in water) for a series of alcohols. The alcohols are grouped according to the number of carbon atoms and are arranged within each group by increasing number of hydroxyl groups.

Alcohol	Condensed structural formula	Partition coefficient
Methanol	CH_3OH	0.14
Ethanol	CH_3CH_2OH	0.26
Ethylene glycol	$HOCH_2CH_2OH$	0.0053
1-Propanol	$CH_3CH_2CH_2OH$	1.9
Propylene glycol	$CH_3CH(OH)CH_2OH$	0.018
Glycerol	$HOCH_2CH(OH)CH_2OH$	0.00066

Data from R. Collander, 1949, *Acta Chem. Scand.* 3:717. (Reprinted with permission from *Acta Chemica Scandinavica*.)

The molecular structure of the permeant is another significant factor in determining its rate of penetration. The condensed structural formulas of the alcohols you will be testing are also given in Table 7.1. An important molecular feature is the presence of hydroxyl groups. Since hydroxyl groups form hydrogen bonds with the water molecules surrounding the cell, a compound with OH groups will have a tendency to remain in the aqueous phase. Molecular size is also a factor; if two compounds have the same lipid solubility but differ in size, it is generally the smaller molecule that enters more rapidly. Small hydrophilic molecules that penetrate rapidly are believed to enter via narrow, protein-lined pores in the plasma membrane.

In some cells, certain molecules penetrate the plasma membrane more rapidly than would be predicted by their molecular structure or lipid solubility. Such *facilitated diffusion*, which is also a passive process, appears to be mediated by specific carrier proteins in the membrane. A number of molecules, especially ions, can enter and leave the cell *against* a concentration gradient. Such *active transport*, as it is termed, requires an expenditure of energy by the cell.

Permeability Study in *Elodea* Leaf Cells. In this permeability study, you will be repeating the classic study of Overton, but with a different plant, *Elodea*. Each alcohol to which the *Elodea* leaf will be exposed is in an *isotonic* solution of sucrose so that the effect of the alcohol alone can be followed. Rapidly penetrating alcohols will establish equal concentrations almost immediately on both sides of the plasma membrane, and there is no visible effect on the cell. If the alcohol enters slowly, the concentration of the alcohol will be higher outside the cell for a while. During this interval, the water, which can penetrate the plasma membrane very rapidly, leaves the cell, and plasmolysis occurs. As the alcohol penetrates the cell, water reenters by osmosis, and deplasmolysis occurs. The time it takes from the initial exposure to the alcohol solution until deplasmolysis begins is a measure of the time required for penetration.

PROCEDURES

PLASMOLYSIS AND DEPLASMOLYSIS

1. Prepare a wet mount of an *Elodea* leaf in a 0.6 M sucrose solution as follows. Remove a leaf from a healthy sprig, place the leaf on a clean slide with the upper leaf surface facing up, drain the excess water, and add two drops of the sucrose solution and a coverslip.

2. With the green filter removed, examine the preparation under high-dry for plasmolysis. Then, on page 85, draw a plasmolyzed cell, labeling the cell wall, outer margin of the cytoplasm (plasma membrane), inner margin of the cytoplasm where it contacts the vacuole membrane (tonoplast), chloroplasts, and nucleus (if visible).

3. From what part of the cell has most of the water been lost? What is now in the space between the plasma membrane and the cell wall? Enter both answers on page 81 and then clean the slide.

4. Prepare wet mounts of individual *Elodea* leaves in 0.2 M, 0.3 M, 0.4 M, and 0.5 M sucrose solutions, following the same procedure as in step 1. All four wet mounts can be prepared at the same time.

5. Examine each preparation under high-dry periodically for 5 min. In each preparation, note whether plasmolysis has occurred and, if so, the extent of plasmolysis. Then answer questions 3–6 on page 81.

6. To observe deplasmolysis, carefully remove the coverslip from a preparation exhibiting plasmolysis, drain the sucrose solution, and add two drops of distilled water. After 1 min, place a coverslip on the preparation. Examine under high-dry for 5 min and then give a brief description of deplasmolysis in item 7 on page 81.

MEASUREMENT OF THE RELATIVE RATES OF PENETRATION OF A SERIES OF ALCOHOLS IN *ELODEA*

The following alcohols are to be tested: methanol, ethanol, ethylene glycol, 1-propanol, propylene glycol, glycerol. The concentration of each alcohol is 0.4 M in a sucrose solution that is approximately isotonic. The alcohol–sucrose solutions should be tested *one at a time* in *Elodea* preparations.

1. Prepare a wet mount of an individual *Elodea* leaf, with the upper surface facing up, in two drops of the alcohol–sucrose solution to be tested. Remember to drain the excess water from the leaf before adding the alcohol–sucrose solution. Record the time that the wet mount is prepared on Data Sheet 7.1

2. Examine the cells under high-dry for the duration and extent of plasmolysis. If plasmolysis does not occur in 5 min, assume that it will not occur, enter *none* for the extent of plasmolysis on Data Sheet 7.1, and proceed testing the next solution. If plasmolysis does occur, record whether it is *slight*, *moderate*, or *severe*, and continue to observe for deplasmolysis.

3. Watch for the first movement of the cell contents toward the cell wall, using a scale division on the ocular micrometer as a reference point. Record the time that deplasmolysis begins on Data Sheet 7.1. If deplasmolysis does not begin in 10 min, assume that it will not occur and note this on the data sheet.

4. Answer questions 8–14 on pages 81 and 82.

REFERENCES

Davson, H. and Danielli, J. F. 1952. *The Permeability of Natural Membranes*. Cambridge Univ. Press. Reprinted in 1970 by Hafner Press, a division of Macmillan, New York.

Finean, J. B., Coleman, R., and Michell, R. H. 1984. *Membranes and Their Cellular Functions*, 3rd ed. Blackwell Scientific Publications, Oxford, England.

Giese, A. C. 1979. *Cell Physiology*, 5th ed., pp. 371–394. W. B. Saunders, Philadelphia.

EXERCISES AND QUESTIONS

1. During plasmolysis, most of the water is lost from the

_____

2. In a plasmolyzed *Elodea* leaf cell, the space between the plasma membrane and the cell wall contains

_____

3. In which concentration(s) of sucrose is plasmolysis *not* observed?

_____

4. In which concentration(s) of sucrose is plasmolysis observed?

_____

5. Is the extent of plasmolysis the same for all hypertonic concentrations or is the effect progressive?

_____

6. Between which two concentrations does the isotonic concentration of sucrose lie?

_____

7. During deplasmolysis, _____

_____

8. For each alcohol on Data Sheet 7.1, calculate the elapsed time until deplasmolysis began.

9. List the alcohols in the order in which they penetrate the plasma membrane of *Elodea* leaf cells, the most rapid at the top of the list.

_____

_____

_____

_____

_____

_____

10. Describe the relationship between the duration of plasmolysis and the extent of plasmolysis.

_____

_____

11. How do your data from the *Elodea* permeability study demonstrate Overton's rule?

_____

_____

12. Compare the penetration rates of the 2-carbon alcohols, ethanol and ethylene glycol. How can the observed difference be explained on the basis of the ether:water partition coefficients? on the basis of molecular structure?

_____

_____

_____

_____

13. Compare the penetration rates of the 3-carbon alcohols, 1-propanol, propylene glycol, and glycerol. How can the observed differences be explained on the basis of the ether:water partition coefficients? on the basis of molecular structure?

_____

_____

_____

_____

14. Is there a measurable difference in the rate of penetration of the two glycols?

_____ If so, give a reason for the observed difference.

_____

_____

_____

82

Data Sheet 7.1

Duration and extent of plasmolysis in *Elodea* leaf cells exposed to 0.4 M solutions of various alcohols in an isotonic sucrose solution.

Alcohol	Time wet mount is prepared	Extent of plasmolysis (none, slight, moderate, severe)	Time deplasmolysis begins	Elapsed time (min) until deplasmolysis begins
Methanol				
Ethanol				
Ethylene glycol				
1-Propanol				
Propylene glycol				
Glycerol				

CELL FRACTIONATION

Cell fractionation is used by the cell biologist to investigate the biochemistry and physiology of organelles outside of the complex environment of the intact cell. The basic methodology used to isolate organelles was developed in the 1930s when A. Claude, R. R. Bensley, and N. L. Hoerr isolated mitochondria and S. Granick isolated chloroplasts. The general procedure in all cell fractionation methods consists of disrupting cell boundaries to release the cell contents and then centrifuging to separate the organelles. The purity of each fraction can be confirmed by microscopic examination or with the use of biochemical markers. In this project, you will use cell fractionation methods to obtain the chloroplast fraction from spinach and the nuclear and mitochondrial fractions from cauliflower. You will analyze the sedimentation behavior of the chloroplasts and investigate the effects of two chemicals on the morphology of chloroplasts in wet mounts.

HOMOGENIZATION

To prepare tissues for cell fractionation, they must first be minced and suspended in an appropriate medium, often a buffered, isotonic sugar or salt solution. To maintain the integrity of the organelles and enzymes during the fractionation procedure, all solutions and containers must be ice-cold. After being minced, the tissue is ready for *homogenization*, a procedure that disrupts cell boundaries and releases the cell contents, undamaged, into the suspension medium. The suspension of cell constituents is termed the *homogenate*. For the homogenization of the spinach and cauliflower cells in this project, you will use a mortar and pestle (Fig. 8.1), with sand added as an abrasive.

CENTRIFUGATION

Because of differences in size and density, each cell component behaves somewhat differently when subjected to a centrifugal force. The force generated by the centrifuge is expressed as a *relative centrifugal force (RCF)* in g units, each g being the force of the earth's gravity exerted upon a mass. The RCF is a function of the speed of centrifugation in *revolutions per minute (rpm)* and the distance of the particle from the axis of rotation. Knowing this distance (x) and the rpm, one can readily determine the RCF from the equation

$$\text{RCF} = 1.119 \times 10^{-5} (\text{rpm})^2 x \tag{8-1}$$

where x is taken as the distance (in centimeters) from the axis of rotation to the outer margin of the centrifuge tube. Most *low-speed clinical centrifuges* have a maximum

1. Tissue is cut up and homogenized in an isotonic aqueous medium.

Mortar and pestle for homogenizing plant tissue

Centrifugation 600 g

Postnuclear supernatant

Nuclear fraction

Decant

2. Homogenate is centrifuged at 600 g for 10 min.

Centrifugation 10,000 g

Postmitochondrial supernatant

Mitochondrial fraction

Decant

3. Postnuclear supernatant is centrifuged at 10,000 g for 30 min.

Centrifugation 105,000 g

Postmicrosomal supernatant (cytosol)

Microsomal fraction

4. Postmitochondrial supernatant is centrifuged at 105,000 g for 60 min.

FIGURE 8.1
Cell fractionation procedure with differential centrifugation and a swinging-bucket rotor.

CELL FRACTIONATION

speed of several thousand rpm. A *high-speed centrifuge*, which often has a refrigerated rotor chamber, can operate at speeds up to 25,000 rpm. An *ultracentrifuge* may have a maximum speed in excess of 75,000 rpm. Its rotor chamber is refrigerated and has a partial vacuum.

The two common types of rotors used in centrifuges are the *swinging-bucket rotor* and the *fixed-angle rotor*. In a swinging-bucket rotor there is a mount that allows the centrifuge tubes to pivot from the vertical to the horizontal position during centrifugation. In a fixed-angle rotor, the centrifuge tubes are held at a specific angle, and the material collects at an angle at the bottom of the tubes. A fixed-angle rotor is more efficient than is a swinging-bucket rotor for sedimenting particles. On the other hand, swinging-bucket rotors are much better suited for *density gradient centrifugation*, which is described shortly.

Differential Centrifugation.   In differential centrifugation, the homogenate is centrifuged repeatedly at successively higher speeds, sedimenting progressively smaller particles. The method is illustrated in Figure 8.1. The first centrifugation is at 600 *g* for 10 minutes, which sediments the *nuclear fraction*. This pellet contains nuclei, intact cells, and tissue debris. For further separation, the *postnuclear supernatant*, or suspended particles and liquid above the nuclear pellet, is transferred to another centrifuge tube and spun at 10,000 *g* for 30 minutes in a refrigerated, high-speed centrifuge. The sedimented material is termed the *mitochondrial fraction*; it contains mitochondria, microbodies, and, from homogenates of animal tissues, also lysosomes. In studies that require the ribosomes or endoplasmic reticulum, which are in the *microsomal fraction*, the *postmitochondrial supernatant* is centrifuged at very high RCF (105,000 *g*), using an ultracentrifuge. The *microsomal supernatant* is the soluble phase of the cell, i.e., the cytosol. It should be noted that the isolated fractions are not pure; there is some contamination in each fraction by smaller organelles that initially were at the bottom of the suspension.

For the isolation of chloroplasts from spinach cells, you will use a modification of a method outlined by F. R. Whatley and D. I. Arnon (1962). The spinach is homogenized in a buffered salt solution. After the plant tissue has been ground up in the mortar, the homogenate is filtered through cheesecloth to remove the larger pieces of tissue. The filtrate is first centrifuged at 200 *g* for 1 minute. The brief centrifugation at low RCF sediments sand, leaf debris, and whole cells. The supernatant is then spun at 1300 *g* for 5 minutes, sedimenting most of the chloroplasts. Nuclei are also in this fraction but comprise only a small portion of the pellet.

To obtain the nuclear and mitochondrial fractions of cauliflower cells, you will use a modification of a procedure for differential centrifugation outlined by W. D. Bonner (1967). The tissue is homogenized in a buffered mannitol solution. After filtration through cheesecloth, the homogenate is centrifuged as illustrated in Figure 8.1, through step 3.

Density Gradient Centrifugation.   For the best resolution of cell fractions and the least contamination by extraneous organelles, another method, density gradient centrifugation, is used. Here, the homogenate is added to the top of a centrifuge tube that contains a sucrose gradient in which the most concentrated (most dense) sucrose solution is at the bottom of the tube and the least concentrated (least dense) sucrose solution is at the top. Figure 8.2 illustrates the results of *rate separation*, one of several density gradient methods. As the tube spins, the organelles distribute themselves in the sucrose gradient by size, the largest organelles sedimenting most rapidly. With a liver homogenate, for example, after centrifugation at 10,000 *g* for 20 minutes, the nuclei are in a layer closest to the bottom of the tube, the mitochondria are in a layer above the nuclei, and the microsomal fraction is still higher in the centrifuge tube. The topmost layer contains the cytosol. Note that all the fractions have been separated with a single round of centrifugation. Following centrifugation, the cell fractions are

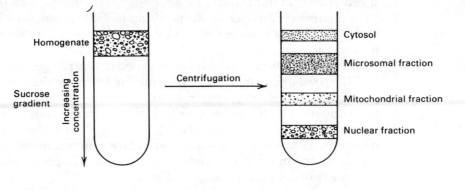

FIGURE 8.2
Cell fractionation with density gradient centrifugation.

removed by puncturing the bottom of the tube and allowing the contents in each layer to drip into a separate test tube.

SEDIMENTATION COEFFICIENT

The *sedimentation coefficient* is a physical constant frequently encountered in biological literature. Many cell structures, especially macromolecules, are often referred to by their sedimentation coefficient. These include proteins, nucleic acids, and small organelles, like the ribosomes. A particle's sedimentation coefficient (s) is determined from its rate of sedimentation under conditions of unit acceleration. Knowing the distance that a particle travels during centrifugation, the duration of the centrifugation, and the rpm, one can calculate s from the equation

$$s = \frac{210 \log (x_2/x_1)}{(\text{rpm})^2 (t_2 - t_1)} \tag{8-2}$$

where x_1 is the distance (in centimeters) from the axis of rotation to the particle at the start of centrifugation, x_2 is the distance from the axis of rotation to the particle at the end of centrifugation, $(t_2 - t_1)$ is the elapsed time (in seconds) of centrifugation, and 210 is a constant. The units for s are seconds.

An example will illustrate the use of Equation 8-2. Assume that a certain particle initially is at the top of the suspension, 7.6 cm from the axis of rotation ($x_1 = 7.6$ cm). If, after centrifugation at 60,000 rpm for 2.5 hours, the particle is 12.5 cm from the axis of rotation ($x_2 = 12.5$ cm), its sedimentation coefficient is

$$s = \frac{210 \log (12.5/7.6)}{(60,000)^2 (2.5) (60) (60)} = 14 \times 10^{-13} \text{ sec}$$

The s values of most cellular macromolecules lie between 1×10^{-13} sec and 200×10^{-13} sec. For convenience, we let "10^{-13} sec" equal a new unit, the *Svedberg unit* (S). Thus, in the example, the sedimentation coefficient of the hypothetical particle is written as 14S. The sedimentation coefficients have been determined for most cellular macromolecules. The s values are usually determined with an *analytical ultracentrifuge*, in which the sedimenting particles can be observed with a special optical system during the centrifugation.

The factors that determine how rapidly a particle sediments in a centrifugal field are the particle's *radius* and its *effective density*, which is the difference between its density and the density of the liquid through which it is moving. The more important factor is the particle's radius so that, in general, the larger the particle, the greater is its sedimentation coefficient. In fact, knowing the sedimentation coefficient of a macromolecule, one can determine the particle's molecular weight. Not surprisingly, the sedimentation coefficients of the larger organelles (mitochondria, chloroplasts, nuclei) are orders of magnitude greater than are the values obtained for macromolecules.

FIGURE 8.3

Horizontal orientation of a centrifuge tube in a swinging-bucket rotor during centrifugation: x_1 = distance from the axis of rotation to the top of the suspension; x_2 = distance from the axis of rotation to the top of the pellet; A = distance from the bottom of the centrifuge tube to the top of the suspension; and B = distance from the bottom of the centrifuge tube to the top of the pellet. Distances x_1 and x_2 can be obtained after centrifugation by measuring A and B; specifically, x_1 = rotor radius − A and x_2 = rotor radius − B.

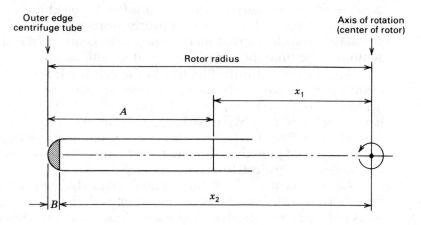

In this project, you will calculate a sedimentation coefficient for spinach chloroplasts from the conditions under which you sediment them. The chloroplast fraction suspended in a cold, buffered salt solution will be centrifuged for 5 min in a swinging-bucket rotor. During the centrifugation, a pellet of chloroplasts will collect at the bottom of the centrifuge tube. After centrifugation, you can calculate an s value for those chloroplasts that moved from the top of the suspension to the top of the pellet. For the calculation of s, you will need to measure the distance from the axis of rotation at the start of centrifugation (x_1) and the distance from the axis of rotation at the end of centrifugation (x_2). These distances are depicted in Figure 8.3.

MICROSCOPIC EXAMINATION OF THE FRACTIONS

All the isolated fractions will be examined microscopically. Chloroplasts are relatively large organelles, so that internal detail is discernible with the light microscope. The electron microscope reveals that the chloroplast has two limiting membranes (Fig. 8.4). Internally, there are *grana* (sing., *granum*), which are stacks of flattened membranous sacs. Each flattened sac is called a *thylakoid*. The matrix surrounding the internal membranes is termed the *stroma*. The thylakoids of different grana are connected by an anastomosing network of membranous channels, the *stroma lamellae*.

After the normal morphology of the spinach chloroplasts has been examined microscopically, you will study the effects of two organic compounds on chloroplast morphology. The chloroplasts in two wet mounts will be treated with a saturated

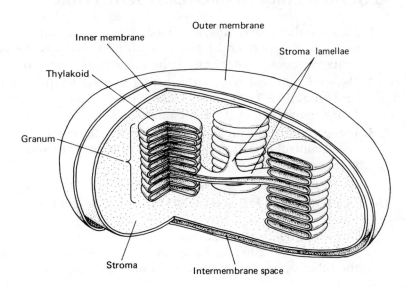

FIGURE 8.4

Three-dimensional representation of a chloroplast as reconstructed from electron micrographs.

aqueous solution of *urea* and a 2% aqueous solution of *Triton X-100*, which is a strong detergent. Urea is known to denature proteins, some of which may thereby be extracted from chloroplast membranes. Detergents, on the other hand, are known to solubilize the lipid bilayer of biological membranes.

The nuclei and mitochondria in the cell fractions of cauliflower cells will be examined microscopically, using two specific stains. Nuclei will be stained green with acidulated methyl green, which is specific for DNA (see Project 3, p. 23). *Nucleoli* (sing., *nucleolus*) are obvious structures within the nuclei since they do not stain with methyl green. Each nucleolus appears as a prominent, round, clear area. Nucleoli are rich in RNA and protein as they are the structures in which the ribosomal subunits are assembled (see Project 16).

Mitochondria will be stained with the vital dye, *Janus green B*. After the staining, mitochondria will appear blue-green, which is the color of the dye in its oxidized form; in its reduced form, the dye is colorless. Presumably, the Janus green B is kept in the oxidized state by the cytochrome oxidase system of mitochondria. Since the mitochondria are very small, you will not be able to resolve any internal structure with the light microscope. A reconstruction based on studies with the electron microscope is presented in Figure 9.1 (p. 103).

PROCEDURES

For all fractionation procedures, you should work in teams of four. For the calculation of the sedimentation coefficient and the microscopic examination of the fractions, work individually.

HOMOGENIZATION OF SPINACH

1. Weigh out 4 g of fresh spinach leaves from which the major veins have been removed.
2. Cut the leaves into small pieces with scissors and then place in a chilled mortar with 15 ml of ice-cold NaCl-buffer and a sprinkling of purified sand. Grind the tissue with a chilled pestle for 2 min.
3. Filter the suspension through four layers of cheesecloth into a chilled 15-ml centrifuge tube. Wring out the juice into the centrifuge tube and then place the tube in an ice-water bath.

DETERMINATION OF THE CENTRIFUGATION SPEEDS REQUIRED FOR CHLOROPLAST ISOLATION

1. Measure the *rotor radius*, i.e., the distance (in centimeters) from the center of the rotor to the bottom of the centrifuge tube, positioned horizontally in the swinging-bucket rotor. Enter the value in the table on page 97.
2. Using Equation 8-1, calculate the rpm required for each of the two centrifugations (200 g and 1300 g) used in obtaining the chloroplast fraction. Enter the values in the table on page 97.
3. Using a centrifuge tachometer, determine the correct settings on the centrifuge's speed control to obtain the required rpm for 200 g and the required rpm for 1300 g.

ISOLATION OF THE CHLOROPLAST FRACTION

1. Remove the centrifuge tube from the ice-water bath and centrifuge the homogenate at 200 g for 1 min. Make sure that the centrifuge tubes are balanced; the tubes opposite each other should have the same total volume.

2. Decant the supernatant into a clean, chilled centrifuge tube and spin at 1300 g for 5 min. Again, be sure that the centrifuge tubes are balanced.

3. After centrifugation, measure distances A and B (see Fig. 8.3). A is the distance (in centimeters) from the bottom of the centrifuge tube to the top of the suspension, and B is the distance from the bottom of the centrifuge tube to the top of the pellet. Enter these values on page 97; you will use them to calculate the sedimentation coefficient.

4. Decant and discard the supernatant and then add 10 ml of ice-cold NaCl-buffer to the pellet in the centrifuge tube. With a Pasteur pipet, thoroughly resuspend the pellet.

5. Cover the centrifuge tube with Parafilm, invert several times, and then place it in an ice-water bath.

DETERMINATION OF THE SEDIMENTATION COEFFICIENT

1. Determine the sedimentation coefficient for the chloroplasts that initially were at the top of the suspension and after centrifugation were at the top of the pellet. To determine s, you will have to calculate the distances x_1 and x_2. As can be seen from Figure 8.3, x_1 = rotor radius − A, and x_2 = rotor radius − B. Enter the values for x_1 and x_2 in the table on page 97.

2. Enter the values for rpm and $(t_2 - t_1)$ (in seconds). Then, using Equation 8-2, calculate s, showing all calculations. Enter the values for s and S in the table.

3. The S values encountered in the literature are corrected so that the sedimentation coefficients are what they would be if the suspending medium were pure water at 20° C. Given that the suspending medium used for the chloroplast isolation in this project is a cold, salt solution, comment (on p. 97) on whether the S value obtained is an underestimate or overestimate of the S value corrected for sedimentation in pure water at 20° C.

4. On page 97, explain why it would have been better to start with all the chloroplasts layered at the top of the suspending medium.

MICROSCOPIC EXAMINATION OF THE CHLOROPLAST FRACTION

1. With a Pasteur pipet, remove a drop of the chloroplast suspension and prepare a wet mount. Use a small drop so that you will not have to press out excess liquid; pressure might damage the chloroplasts.

2. Examine the chloroplasts under high-dry and oil-immersion, with the green filter removed.

3. Measure the length and width of five chloroplasts. Enter these values and the means on page 98.

4. Observe the morphology of several chloroplasts. Look for some in which the interior is not homogeneous and then answer the question on page 98.

5. On the blank paper provided at the end of this project (p. 101), draw a chloroplast that shows internal detail, labeling as many structures as possible.

MICROSCOPIC EXAMINATION OF UREA-TREATED CHLOROPLASTS

1. Place a small drop of the chloroplast suspension on a clean slide. Add a small drop of the saturated urea solution and a coverslip. Again, use small drops so that you will not have to press out excess liquid.

2. Examine the chloroplasts under high-dry and oil-immersion for about 5 min and then answer the questions on page 98.

3. Draw a chloroplast that shows the altered morphology.

MICROSCOPIC EXAMINATION OF TRITON-TREATED CHLOROPLASTS

1. Prepare another wet mount, this time using a small drop of the chloroplast suspension and a small drop of the 2% Triton X-100 solution.

2. Examine the preparation under high-dry and oil-immersion and then answer the questions on page 98.

ISOLATION OF THE NUCLEAR AND MITOCHONDRIAL FRACTIONS

1. Using a single-edge razor blade, remove a total of 20 g of the outer 2–3 mm of the cauliflower surface.

2. Place the tissue in a chilled mortar with 40 ml of ice-cold *mannitol grinding medium* and 5 g of cold, purified sand. Grind the tissue with a chilled pestle for 4 min.

3. Filter the suspension through four layers of cheesecloth into a chilled 50-ml centrifuge tube and also wring out the juice into the tube.

4. Centrifuge the filtrate (homogenate) at 600 g for 10 min at 0°–4° C. Make sure that the centrifuge tubes are balanced.

5. Decant the postnuclear supernatant into a clean, chilled centrifuge tube(s) and place the centrifuge tube with the nuclear pellet in an ice-water bath.

6. Centrifuge the postnuclear supernatant at 10,000 g for 30 min at 0°–4° C. Again, be sure that the centrifuge tubes are balanced. During the centrifugation, the nuclear pellet can be examined microscopically, as described below.

7. Decant and discard the postmitochondrial supernatant and add 5 ml of the ice-cold mannitol medium to the mitochondrial pellet.

8. With a spatula, scrape the mitochondrial pellet from the wall of the centrifuge tube, and then with a Pasteur pipet, thoroughly resuspend the sediment in the mannitol medium.

9. Transfer the mitochondrial suspension to a test tube and place it in an ice-water bath.

MICROSCOPIC EXAMINATION OF THE NUCLEAR FRACTION

1. With a spatula, remove a tiny amount of the nuclear pellet and smear the material on a clean slide. Immediately, before the smear dries, add several drops of methyl green stain. After 5 min, add a coverslip and gently press out the excess stain with a paper towel.

2. Examine the preparation under high-dry, with the green filter removed. When you have located a region with nuclei, switch to the oil-immersion objective. For each of five round nuclei, observe the number of nucleoli and measure the nuclear diameter. Enter the values on page 98.

3. Calculate the mean diameter of the nuclei and answer the question on nucleoli on page 98.

4. On the blank paper provided (p. 101), draw a typical nucleus, labeling each nucleolus.

MICROSCOPIC EXAMINATION OF THE MITOCHONDRIAL FRACTION

1. Place five drops of the mitochondrial suspension in a small test tube. Add five drops of the Janus green B stain, swirl the tube, and allow the mixture to remain at room temperature for 5 min.
2. Place a drop of the mixture on a clean slide, add a coverslip, and examine under high-dry and oil-immersion, with the green filter removed. Identify the mitochondria, which appear as very small, blue-green structures.
3. Measure the length of five mitochondria. Enter the five values and the mean on page 99.
4. Observe the morphology of the mitochondria and then answer the question on page 99.

REFERENCES

Bonner, Jr., W. D. 1967. A general method for the preparation of plant mitochondria. In *Methods in Enzymology*, Vol. 10, Estabrook, R. W. and Pullman, M. E., eds., pp. 126–133. Academic Press, New York.

Sheeler, P. 1981. *Centrifugation in Biology and Medical Science*, pp. 10–94. John Wiley, New York.

Whatley, F. R. and Arnon, D. I. 1962. Photosynthetic phosphorylation in plants. In *Methods in Enzymology*, Vol. 6, Colowick, S. P. and Kaplan, N. O., eds., pp. 308–313. Academic Press, New York.

EXERCISES AND QUESTIONS

DETERMINATION OF THE CENTRIFUGATION SPEEDS REQUIRED FOR CHLOROPLAST ISOLATION

RCF	Rotor radius (cm)	Required rpm
200 g		
1300 g		

ISOLATION OF THE CHLOROPLAST FRACTION

$$A = \text{\_\_\_\_\_} \text{ cm}; B = \text{\_\_\_\_\_} \text{ cm}$$

DETERMINATION OF THE SEDIMENTATION COEFFICIENT

x_1	x_2	rpm	$(t_2 - t_1)$	s	S

Compared to the S value for chloroplasts sedimented in pure water at 20° C, is the S value obtained in the NaCl-buffer an underestimate or overestimate? Explain.

_____

_____

Why would it have been better to start with all the chloroplasts layered at the top of the suspending medium?

_____

_____

_____

_____

OBSERVATIONS AND QUESTIONS

MICROSCOPIC EXAMINATION OF THE CHLOROPLAST FRACTION

Chloroplast length and width (μm)

Chloroplast	1	2	3	4	5	
Length	_____	_____	_____	_____	_____	$\overline{X}$ = _____ μm
Width	_____	_____	_____	_____	_____	$\overline{X}$ = _____ μm

What structures are visible within the chloroplasts?

_____

MICROSCOPIC EXAMINATION OF UREA-TREATED CHLOROPLASTS

What morphological changes are occurring in the chloroplasts?

_____

_____

In light of what you know about the action of urea, explain your observations.

_____

_____

MICROSCOPIC EXAMINATION OF TRITON-TREATED CHLOROPLASTS

What morphological changes have occurred in the chloroplasts?

_____

In light of what you know about the action of Triton X-100, explain your observations.

_____

_____

MICROSCOPIC EXAMINATION OF THE NUCLEAR FRACTION

Nucleus	1	2	3	4	5	
Number of nucleoli	_____	_____	_____	_____	_____	
Nuclear diameter (μm)	_____	_____	_____	_____	_____	$\overline{X}$ = _____ μm

How many nucleoli are visible in most nuclei? _____

MICROSCOPIC EXAMINATION OF THE MITOCHONDRIAL FRACTION

Mitochondrial length (μm)

_____   _____   _____   _____   _____ $\overline{X}$ = _____ μm

Do all mitochondria have the same shape? _____ Describe their morphology.

_____

SUCCINATE DEHYDROGENASE ACTIVITY OF MITOCHONDRIA

INTRODUCTION

Contained within the mitochondria is the biochemical machinery for *cellular respiration*, the aerobic process by which sugars, fatty acids, and amino acids are broken down to carbon dioxide and water and their chemical energy captured as *adenosine triphosphate* (*ATP*). A key series of reactions in this aerobic process is the *tricarboxylic acid* (*Krebs*) *cycle*, a complex pathway involving many enzymes and metabolic intermediates. In this project, you will measure the activity of one of the best studied enzymes in the Krebs cycle, *succinate dehydrogenase*, which catalyzes the oxidation of succinate to fumarate. The *in vitro* analysis will be carried out with the mitochondrial fraction, isolated from the cells of cauliflower.

ENZYMES OF THE MITOCHONDRIA

The morphology of mitochondria can vary from almost spherical to fibrillar depending on cell type, but typically they are sausage-shaped organelles, 0.2–1.0 μm in diameter and 1–4 μm in length. The structural features of a typical mitochondrion are diagrammed in Figure 9.1. There are two membrane systems, an *outer membrane* that separates the mitochondrial contents from the cytoplasm, and a convoluted *inner membrane*, the folds of which are termed *cristae*. The central region surrounding the cristae contains a gel-like *matrix*.

FIGURE 9.1
Three-dimensional representation of a mitochondrion as reconstructed from electron micrographs.

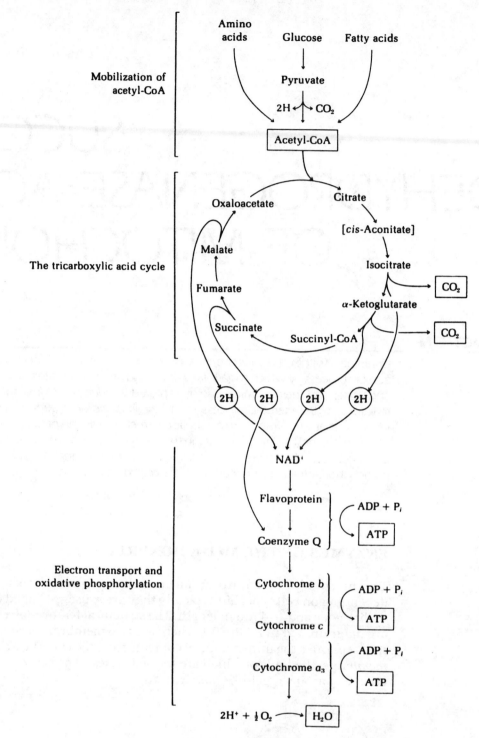

FIGURE 9.2
Summary of the reactions
in cellular respiration.
(From A. L. Lehninger, 1975,
Biochemistry, 2nd ed., p. 445,
Worth Publishers, Inc., New
York.)

The reactions in cellular respiration are shown in Figure 9.2. The reactions from pyruvate through *electron transport* have been localized in the various membranes and compartments of the mitochondrion. The cytochromes and other electron/hydrogen carriers are located in the inner membranes. It is during the process of electron transport that most of the ATP is generated. All but one of the enzymes of the tricarboxylic acid cycle are located in the mitochondrial matrix. The one exception is succinate dehydrogenase and its coenzyme, *flavin adenine dinucleotide (FAD)*. This enzyme complex, which can be represented by *E-FAD*, is tightly bound to the inner membranes of the mitochondrion. The function of the succinate dehydrogenase–FAD complex is the oxidation of succinate to fumarate:

$$\text{succinate} \quad + \quad \text{E-FAD} \quad \longrightarrow \quad \text{fumarate} \quad + \quad \text{E-FADH}_2$$

As can be seen from this equation, the hydrogens removed from the succinate are carried by FAD. The hydrogens in the reduced complex (E-FADH$_2$) are eventually transferred to coenzyme Q, a component of the electron transport chain (Fig. 9.2).

MEASUREMENT OF SUCCINATE DEHYDROGENASE ACTIVITY

In this project, you will be measuring succinate dehydrogenase activity in isolated mitochondria. The succinate $\longrightarrow$ fumarate reaction is measured by monitoring the reduction of an *artificial* electron acceptor. To use an artificial electron acceptor, the normal path of electrons in the electron transport chain must be blocked. This is accomplished by adding either potassium cyanide or sodium azide to the reaction mixture. These poisons inhibit the transfer of electrons from cytochrome a$_3$ to the final acceptor, oxygen, so that electrons cannot be passed along by the preceding cytochromes and coenzyme Q. Instead, they can be picked up by an artificial electron acceptor such as *2,6-dichlorophenolindophenol* (*DCIP*). The reduction of DCIP by E-FADH$_2$ can be measured spectrophotometrically since the oxidized form of the dye is blue and the reduced form is colorless. The reaction can be summarized:

$$\text{E-FADH}_2 \quad + \quad \underset{\text{blue}}{\text{DCIP}_{\text{ox}}} \longrightarrow \text{E-FAD} \quad + \quad \underset{\text{colorless}}{\text{DCIP}_{\text{re}}}$$

Thus, the change in absorbance, measured at 600 nm, can be used to measure the extent of the reaction.

ENZYME KINETICS

Enzymes are catalysts, accelerating the rates of reactions so that they occur very rapidly under conditions in the cell. We are interested in factors that affect the rate behavior, or *kinetics*, of enzyme-catalyzed reactions. Much of the pioneering work in the area of enzyme kinetics was done by L. Michaelis and M. L. Menten (1913), who formulated a theory that describes the initial events in enzyme catalysis. The theory assumes that the enzyme (*E*) and substrate (*S*) combine reversibly to form an enzyme–substrate (*ES*) complex. The ES complex then breaks down to form free enzyme and product (*P*). These reactions are summarized by the equation

$$E \; + \; S \underset{k_2}{\overset{k_1}{\rightleftharpoons}} ES \overset{k_3}{\longrightarrow} E \; + \; P$$

where k_1, k_2, and k_3 are the rate constants. Though the derivation is not presented here, it can be shown that the initial velocity (v_0) of an enzyme-catalyzed reaction is given by the equation

$$v_0 \; = \; \frac{k_3\,[E][S]}{\left(\dfrac{k_2 \; + \; k_3}{k_1}\right) \; + \; [S]} \tag{9-1}$$

We are concerned with the initial velocity and its relationship to enzyme concentration, so we can work with Equation 9-1, which is a form of the

Michaelis–Menten equation. Since the concentration of substrate in the reaction mixture is known and since k_1, k_2, and k_3 are constants, the equation states that the initial velocity is directly proportional to enzyme concentration.

Effect of Enzyme Concentration.   In this project, you will measure the initial velocity of the succinate $\longrightarrow$ fumarate reaction at several concentrations of the enzyme. To vary the concentration of the enzyme in the reaction mixture, you will add different volumes of the resuspended mitochondrial fraction. Since the total volume in each reaction mixture is the same, the relative *volumes* of the added mitochondrial suspension represent the relative *concentrations* of the enzyme in the mixture. Each reaction mixture contains a very high concentration of substrate so that there will still be a relatively high substrate concentration during the first 5 or 10 min of the assay.

The velocity of the reaction will be monitored by measuring the change in absorbance (ΔA_{600}) with time. Recall that ΔA_{600} shows the extent to which the dye DCIP is reduced by E-FADH$_2$. For each of three enzyme concentrations, you will plot ΔA versus time and, from the initial slope of each curve, determine the initial velocity. The initial velocities can then be plotted against enzyme concentration.

Effect of a Competitive Inhibitor.   The velocity of an enzymatic reaction can be altered by the presence of an inhibitor. You will study the effect of a *competitive inhibitor*, which, as the name suggests, competes with the normal substrate for the active site on the enzyme. *Malonate*, whose molecular structure ($^-OOCCH_2COO^-$) is similar to that of succinate, is a classic example of a competitive inhibitor. Though the malonate binds to the succinate dehydrogenase, it cannot be dehydrogenated. Along with the measurement of v_0 at three enzyme concentrations, you will determine the velocity in the presence of malonate.

Controls.   You will run three controls. In one control, the reaction mixture contains no azide. In a second control there is no added succinate. The third control contains inactivated enzyme. The enzyme is inactivated by heating the mitochondrial suspension, thereby denaturing the protein.

ISOLATION OF MITOCHONDRIA

You will isolate the mitochondrial fraction from cauliflower cells using the same method of differential centrifugation as in Project 8 (Bonner, 1967). You will homogenize the cauliflower tissue in a mortar and pestle (Fig. 8.1, p. 88) with a buffered, isotonic mannitol solution and a quantity of sand. After squeezing the homogenate through cheesecloth to remove the larger pieces, you will centrifuge the filtrate at 600 g for 10 min to sediment the nuclear fraction. You will then centrifuge the supernatant at 10,000 g for 30 min to sediment the mitochondrial fraction. For all cell fractionation procedures, the solutions and containers must be ice-cold.

PROCEDURES

For all procedures, you should work in teams of four.

ISOLATION OF THE MITOCHONDRIAL FRACTION

1. Using a single-edge razor blade, remove a total of 20 g of the outer 2–3 mm of the cauliflower surface.
2. Place the tissue in a chilled mortar with 40 ml of ice-cold *mannitol grinding medium* and 5 g of cold, purified sand. Grind the tissue with a chilled pestle for 4 min.

3. Filter the suspension through four layers of cheesecloth into a chilled 50-ml centrifuge tube; also wring out the juice into the tube.
4. Centrifuge the filtrate at 600 g for 10 min at 0°–4° C. Make sure that the centrifuge tubes are balanced; the tubes opposite each other should have the same total volume.
5. Decant the postnuclear supernatant into a clean, chilled centrifuge tube(s) and spin at 10,000 g for 30 min at 0°–4° C. Again, be sure that the centrifuge tubes are balanced. The nuclear pellet can be discarded.
6. Decant and discard the postmitochondrial supernatant and add 7.0 ml of ice-cold *mannitol assay medium** to the mitochondrial pellet.
7. With a spatula, scrape the mitochondrial pellet from the wall of the centrifuge tube and then with a Pasteur pipet thoroughly resuspend the sediment in the assay medium. It is important that the clumps be completely dispersed.
8. Transfer the mitochondrial suspension to a test tube and place it in an ice-water bath, in which it should be kept during the entire experiment.

MEASUREMENT OF SUCCINATE DEHYDROGENASE ACTIVITY

1. Allow the spectrophotometer to warm up for at least 5 min. Instructions for its use are given in Project 4, pages 30 and 35 (steps 8 and 9 at the bottom). Set the wavelength at 600 nm.
2. Label 10 cuvettes as shown in the table following step 4. Except for the ice-cold mitochondrial suspension, all solutions should be at room temperature.
3. Prepare the mitochondrial suspension for tube 7 by heating a 0.6-ml aliquot in a boiling-water bath for 5 min and then cooling in an ice-water bath.
4. To all cuvettes, add the various solutions given across the top of the table, except for the mitochondrial suspension. First, add the correct volume of assay medium* to all tubes. Then, in the same manner, add the correct volumes of azide,[†] DCIP,* malonate,* and succinate,* as indicated in the table. Cover each cuvette with Parafilm and invert twice to mix the contents.

Tube	Assay Medium*	Azide[†] (0.04 M)	DCIP* (5 × 10^{-4} M)	Malonate* (0.2 M)	Succinate* (0.2 M)	Mitochondrial suspension*
Blank 1	3.7 ml	0.5 ml	—	—	0.5 ml	0.3 ml
1	3.2 ml	0.5 ml	0.5 ml	—	0.5 ml	0.3 ml
Blank 2	3.1 ml	0.5 ml	—	—	0.5 ml	0.9 ml
2	2.6 ml	0.5 ml	0.5 ml	—	0.5 ml	0.9 ml
Blank 3	3.4 ml	0.5 ml	—	—	0.5 ml	0.6 ml
3	2.9 ml	0.5 ml	0.5 ml	—	0.5 ml	0.6 ml
4	2.7 ml	0.5 ml	0.5 ml	0.2 ml	0.5 ml	0.6 ml
5	3.4 ml	—	0.5 ml	—	0.5 ml	0.6 ml
6	3.4 ml	0.5 ml	0.5 ml	—	—	0.6 ml
7	2.9 ml	0.5 ml	0.5 ml	—	0.5 ml	0.6 ml[‡]

*CAUTION: *Do not pipet by mouth.*
[†]CAUTION: *Azide is poisonous; do not pipet by mouth.*
[‡]The mitochondrial suspension to be added to tube 7 will be heated in a boiling-water bath for 5 min.

5. Thoroughly resuspend the mitochondrial suspension with a Pasteur pipet. Then add the correct volume of mitochondrial suspension* to each cuvette, recording the time on Data Sheet 9.1. As soon as the mitochondrial suspension has been

*CAUTION: *Do not pipet by mouth.*
[†]CAUTION: *Azide is poisonous; do not pipet by mouth.*

added, cover the cuvette with Parafilm and invert twice to mix the contents. Remove the Parafilm and set the tube in the test tube rack.

6. When 5 min have elapsed from the time that the mitochondrial suspension was added to tube 1, adjust the spectrophotometer for blank 1 and take the 5-min absorbance reading for tube 1. Adjust for blank 2 and take the 5-min absorbance reading for tube 2. Adjust for blank 3 and take the 5-min absorbance readings for tubes 3–7. Enter all absorbance readings on Data Sheet 9.1

7. At 5-min intervals, measure the absorbance of the seven tubes for 45 min. Always remember to adjust for the three blanks as in step 6. Enter the absorbance readings on Data Sheet 9.1

REFERENCES

Bonner, Jr., W. D. 1967. A general method for the preparation of plant mitochondria. In *Methods in Enzymology*, Vol. 10, Estabrook, R. W. and Pullman, M. E., eds., pp. 126–133. Academic Press, New York.

Lehninger, A. L. 1982. *Principles of Biochemistry*, pp. 207–243, 435–506. Worth, New York.

Sheeler, P. and Bianchi, D. E. 1987. *Cell and Molecular Biology*, 3rd ed., pp. 377–416. John Wiley, New York.

EXERCISES AND QUESTIONS

1. On Data Sheet 9.2, enter for each tube the *total change in absorbance* (ΔA) at each time interval. The total change in absorbance is the difference between the 5-min reading for tube 7 and the reading for each tube at the specified time. Why is the 5-min absorbance reading for tube 7 taken as the 0-min reading for all tubes?

_____

_____

2. On Graph 9.1, plot the reaction rates for tubes 1–4. Plot the total change in absorbance (ordinate) versus time (abscissa). Include the origin as a point for all curves, i.e., the 0-min value for ΔA is 0. Draw the best-fit curve with a ruler and French curve template and label each plot.

3. What volume of mitochondrial suspension gives the highest initial velocity?

_____ ml With what volume of mitochondrial suspension does the velocity

level off most rapidly? _____ ml How do you explain these observations?

_____

_____

_____

4. On Graph 9.2, plot the initial velocity (ordinate) versus enzyme concentration (abscissa) for tubes 1–3. For the initial velocity, take the total change in absorbance after 5 min (from Graph 9.1) and divide by 5 to obtain the ΔA/min. For the different enzyme concentrations, simply use the three volumes of mitochondrial suspension added: 0.3, 0.6, 0.9. A fourth point should pass through the origin. Draw the best-fit line and then comment on what happens to v_0 when the enzyme concentration is doubled and tripled.

_____

_____

5. How does the initial velocity in the reaction mixture with malonate (tube 4) differ from the initial velocity in the comparable mixture without malonate

(tube 3)? _____
Explain.

_____

_____

_____

6. From the data, what is the evidence that the electron transport system is functioning in the isolated mitochondria?

_____

_____

_____

7. From the data, what is the evidence that there is very little, if any, succinate in the mitochondrial fraction?

_____

_____

_____

Data Sheet 9.1

Absorbance readings (A_{600}) at each time interval.

Tube	Time that mitochondrial suspension was added	Time (min)								
		5	10	15	20	25	30	35	40	45
1										
2										
3										
4										
5										
6										
7										

Data Sheet 9.2
Total change in absorbance (ΔA) at each time interval.

Tube	5	10	15	20	25	30	35	40	45
1									
2									
3									
4									
5									
6									
7									

Time (min)

Graph 9.1

Plot of total change in absorbance (ΔA) versus time (min) for tubes 1–4.

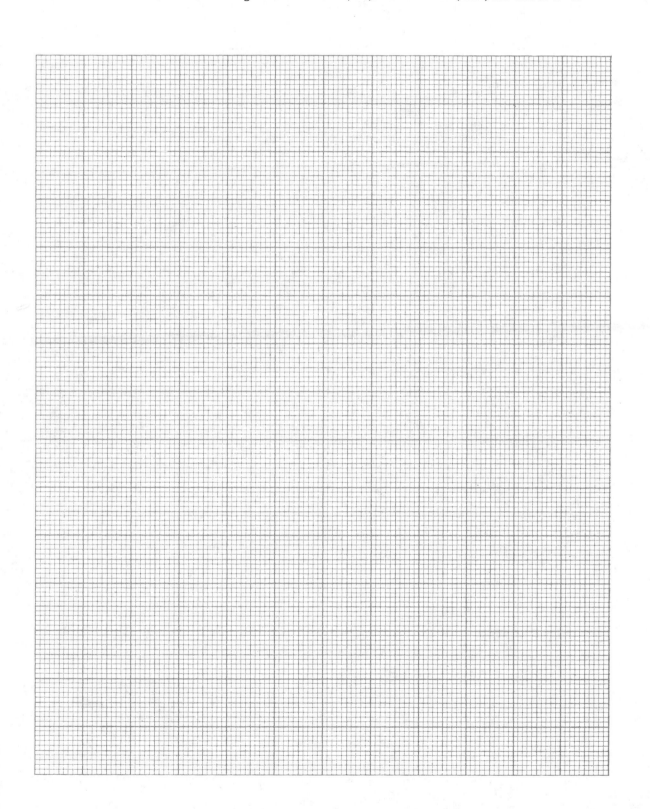

Graph 9.2
Plot of initial velocity (v_0) versus relative enzyme concentration for tubes 1–3.

CHROMATOGRAPHY OF PHOTOSYNTHETIC PIGMENTS

INTRODUCTION

The light energy required for photosynthesis is absorbed and trapped by the photosynthetic pigments. The major pigments of photosynthesis are the *chlorophylls*. The two chlorophylls found in green plants are *chlorophyll a* (*chl a*) and *chlorophyll b* (*chl b*). Light energy is also absorbed by *carotenoids* and *phycobilins*, sometimes referred to as *accessory pigments*. Carotenoids occur in all photosynthetic organisms, while phycobilins occur in the red and blue-green algae. In this project, you will extract the photosynthetic pigments from spinach leaves and separate them using the technique of *paper chromatography*. After separating the pigments, you will obtain their absorption spectra and determine the relative amounts of chlorophylls *a* and *b* in spinach.

PHOTOSYNTHETIC PIGMENTS

Chlorophylls.   The chlorophylls have a similar molecular structure as shown by the formulas of chlorophyll *a* and chlorophyll *b* in Figure 10.1. Each has a *porphyrin ring* and a long *phytol chain*. The porphyrin ring resembles the prosthetic group of hemoglobin and cytochrome but has a central magnesium atom instead of iron. The alternating double and single bonds of the porphyrin ring make chlorophyll an efficient light-absorbing molecule and determine the general shape of the absorption spectrum. Chlorophylls absorb strongly in the red and blue regions of the visible spectrum and transmit in the green. The phytol chain, which is almost devoid of double bonds, contributes little to the absorption spectrum. As is the case for other compounds, the specific absorption maxima of any chlorphyll depends on the solvent in which it is dissolved.

Carotenoids.   There are two classes of carotenoids, the *carotenes* and the *xanthophylls*. Figure 10.2 shows the structural formulas of three common carotenoids in spinach leaves. One is a carotene, *β-carotene*, and the other two are xanthophylls, *lutein* and *violaxanthin*. Another common xanthophyll is *neoxanthin*, not depicted in the figure. All carotenoids have long isoprenoid chains, with alternating double and

R (chl a) = CH_3
R (chl b) = CHO

Porphyrin ring

Phytol chain

FIGURE 10.1
Structural formulas of chlorophyll a (chl a) and chlorophyll b (chl b).

β-Carotene

Lutein

Violaxanthin

FIGURE 10.2
Structural formulas of three common carotenoids: β-carotene, lutein, and violaxanthin.

single bonds. Structurally, the carotenes are composed entirely of carbon and hydrogen, whereas the xanthophylls also contain oxygen. Primarily as a result of their conjugated double bonds, the accessory pigments absorb blue wavelengths of light and transmit yellow and orange light. Some of the absorbed light energy can be transferred to chlorophyll *a*, which participates directly in the conversion of light energy to chemical energy. Since the absorption spectra of the accessory pigments differ from those of the chlorophylls, the effect of accessory pigments is to broaden the range of wavelengths that can be utilized in photosynthesis. Carotenoids also protect the chlorophylls from photooxidation by filtering out harmful wavelengths and by serving preferentially as photooxidizable substrates.

PAPER CHROMATOGRAPHY

Chromatography is a technique used to separate the components of a mixture. There are various types of chromatography (column, paper, thin-layer, gas), but in all cases the separation is achieved by the distribution of components between a fixed or *stationary phase* and a moving or *mobile phase*. In paper chromatography, the components of a mixture are separable into discrete zones on a sheet of filter paper. The mixture is initially spotted or streaked near one end of the paper. If the separated substances are to be extracted later for further analysis, the procedure is called *preparative paper chromatography* (Fig. 10.3). With a capillary tube, the mixture is streaked on the chromatography paper (Fig. 10.3a); enough sample is applied so that

FIGURE 10.3
Preparative paper chromatography. (a) Application of the mixture onto chromatography paper. (b) Chromatography jar with streaked chromatography paper. (c) Chromatogram.

there will be an adequate amount for subsequent extraction and spectrophotometric analysis. For *ascending paper chromatography*, which is depicted in Figure 10.3, the appropriate solvent is added to the bottom of a chromatography jar. The atmosphere in the jar should be saturated with solvent vapor prior to adding the paper. The paper is suspended in the chromatography jar so that the streak is above the level of the solvent (Fig. 10.3b). Then, the solvent moves up the paper by capillary action, past the sample, toward the end of the paper. During this process, termed *development*, the solutes separate and form a trail of discrete bands on the *chromatogram* (Fig. 10.3c).

Mechanism of Separation. In paper chromatography, the stationary phase consists primarily of the water molecules bound to the paper. Chromatography paper contains 6–10% water adsorbed to the cellulose fibers. The solutes in the mixture dissolve to some extent in this bound water. A less significant factor may be the adsorption of the solutes to the cellulose fibers per se. The mobile phase is the moving solvent, in which each solute dissolves to a different extent. Thus, there are two opposing forces. The *driving force*, or moving solvent, tends to carry each solute with it toward the end of the paper. The *resisting force* is due largely to the dissolution of each solute in the cellulose-associated water. It is the interaction of these forces on each component of the mixture that results in the separation of the various solutes.

The degree of separation obtained is largely a function of the relative solubilities of each solute in the stationary (water) and mobile (solvent) phases. Each solute has a characteristic *water:solvent partition coefficient*, defined as (solubility in water phase)/(solubility in solvent phase). During development, each solute is repeatedly partitioned or distributed between the two phases, a process termed *countercurrent distribution*. Figure 10.4 illustrates how countercurrent distribution effects the

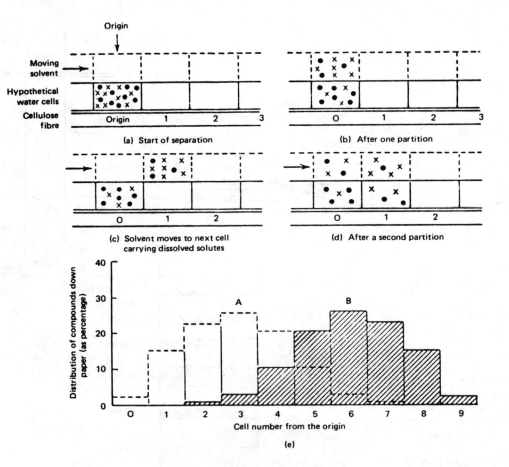

FIGURE 10.4
The process of countercurrent distribution in the separation of solutes A (•) and B (x) by paper chromatography. See text for explanation. (From D. Abbott and R. S. Andrews, 1965, *An Introduction to Chromatography*, p. 4, Houghton Mifflin, Boston. Reproduced by copyright permission of Longman Group Ltd., Harlow Essex, England.)

CHROMATOGRAPHY OF PHOTOSYNTHETIC PIGMENTS

chromatographic separation of a mixture of two solutes, A and B. In this example, A and B have partition coefficients of 2.0 and 0.5, respectively, A being more water-soluble than B. The water associated with the cellulose fibers can be envisioned as occurring in minute "cells." At the *origin*, or point of application, there is initially an equal number of A(•) and B(x) molecules (Fig. 10.4a). As the solvent passes the origin (Fig. 10.4b), each solute distributes itself between the water and solvent according to its partition coefficient (2:1 for A, 1:2 for B). The solvent, now containing some of the mixture, moves over the next cell (Fig. 10.4c). There is now a partition of A and B in cell 1 and a new partition in cell 0 as fresh solvent moves over the origin. As can be seen in Figure 10.4d, there is a tendency for solute B(x) to advance more readily than solute A(•). As the histogram in Figure 10.4e indicates, even 10 such partitions produce a marked separation of A and B. In actuality there are countless cells on a sheet of chromatography paper so that a complete separation can usually be achieved with an appropriate solvent system.

Separation of the components is usually measured by the R_f value. The R_f value is given by the equation

$$R_f = \frac{\text{distance traveled by the solute}}{\text{distance traveled by the solvent from the origin}}$$

For the numerator, the distance is measured from the origin either to the center or to the leading edge of each spot or band. The denominator is the distance from the origin to the solvent front. R_f values can be used to identify the various solutes when the experimental conditions are very carefully controlled.

ANALYSIS OF SPINACH PIGMENTS

The photosynthetic pigments are extracted from spinach by grinding the leaves in acetone. The chromatography paper is then streaked with the spinach extract and suspended in a chromatography jar previously equilibrated with vapors of the solvent, a 9:1 mixture of petroleum ether:acetone. When separation is complete, you will identify the pigment bands by their colors and relative positions on the chromatogram. The major pigments appear in five bands: in order, from the origin to the solvent front, they are chl *b* (olive-green), chl *a* (blue-green), violaxanthin (yellow), lutein (yellow), and, near the leading edge of the solvent front, β-carotene (yellow-orange). Neoxanthin is also present, but its R_f value in this solvent system is close to that of chl *b*, which masks it. Each pigment or pigment group will be eluted from the chromatogram by cutting out each band and soaking the strips of paper in acetone. Violaxanthin and lutein, the two xanthophylls, will be combined and treated as a single group. The absorption spectrum of each pigment or pigment group will then be determined. (The effect of neoxanthin on the absorption spectrum of chl *b* is negligible and can be ignored for this study.)

For the two major pigments, chl *a* and chl *b*, you will perform a quantitative analysis. The absorption coefficients (α) for chl *a* and chl *b* in acetone were determined by G. Mackinney in 1940; for chl *a*, $\alpha_{663} = 80.17$ cm^2/mg, and for chl *b*, $\alpha_{645} = 50.93$ cm^2/mg (663 nm and 645 nm were chosen since they are *absorption peaks* for chl *a* and chl *b*, respectively). Using the Beer–Lambert equation, $A = \alpha cl$ (see Project 4, p. 31), you will determine the concentration (mg/ml) of each chlophyll in the eluted sample. Finally, you will calculate the ratio [chl *a*]/[chl *b*], a value that is characteristic for each plant species.

PROCEDURES

PREPARATION OF THE CHROMATOGRAPHY JAR

For the entire project, you should work in pairs. The solvent mixture is *extremely flammable* so that the chromatography should be carried out in a fume hood. Before handling the chromatography paper, hands should be washed and thoroughly dried. The paper should be handled as little as possible and only by the edges.

1. Cut a strip of Whatman 3MM chromatography paper 7.5 cm wide. The length of the strip should be such that it almost touches the bottom of the chromatography jar when hung from the lid (Fig. 10.3b).
2. Add the solvent mixture to the chromatography jar (in a fume hood) at least 30 min before adding the streaked chromatography paper so that the atmosphere in the jar will become saturated with solvent vapor. Add freshly prepared solvent (9 vol petroleum ether*:1 vol acetone, mixed well) to a height of approximately 2 cm and cover the jar with a lid that is sealed at the rim with Vaseline. Be careful not to get any solvent on the Vaseline.

PREPARATION OF THE LEAF EXTRACT

1. Weigh out 3 g of fresh spinach leaves from which the major veins have been removed.
2. Cut the leaves into small pieces with scissors and then place in a chilled mortar with 15 ml ice-cold acetone and a sprinkling of clean sand. Grind the leaves thoroughly for about 1 min with a chilled pestle.
3. Transfer the liquid and pulp to a 50-ml stoppered test tube. Shake the tube vigorously for 10 sec and then place it in the refrigerator for 10 min. During this time, the white spinach pulp will settle.
4. Using a Pasteur pipet, transfer a portion of the extract, the dark upper layer containing the pigments, to a small stoppered test tube. The pigment extract will be dispensed from this small test tube to minimize evaporation.

PREPARATIVE PAPER CHROMATOGRAPHY

1. With a pencil and ruler, draw a line across the width of the chromatography paper, about 3 cm from the bottom. The extract, which will be streaked on this line, must not be immersed in the solvent.
2. Using a capillary tube, make 10 streak applications of the pigment extract along the line (Fig. 10.3a). The capillary tube is filled by immersing the tip in the extract. The flow from the capillary tube is controlled by finger pressure at the top. Allow each application to air dry before making the next. The final thickness of the streak should be no more than 6–7 mm.
3. Attach the streaked chromatography paper to the hook on the lid of the equilibrated chromatography jar (Fig. 10.3b), being careful not to get any Vaseline on the paper. Allow the chromatogram to develop in the dark or in very dim light for 30–45 min or until there is clean separation of the five bands. Stop the development before the solvent front reaches the end of the paper.
4. Remove the chromatogram from the jar and hold it by the corner until dry. Place the dry chromatogram beneath the chart on Data Sheet 10.1. Next to the chromatogram, mark off in pencil directly on the data sheet the location of the

***CAUTION:** Petroleum ether is extremely flammable and should be kept away from heat, sparks, or an open flame.*

CHROMATOGRAPHY OF PHOTOSYNTHETIC PIGMENTS

origin, solvent front, and the approximate center of each band. Also record the color of each band and what pigment it contains. This must be done before elution.

ELUTION AND SPECTROPHOTOMETRY

1. Allow the spectrophotometer to warm up for at least 5 min. Instructions for its use are given in Project 4, pages 30 and 35 (steps 8 and 9 at the bottom). Set the wavelength at 400 nm.
2. Label five cuvettes as follows: chl *b*, chl *a*, xanthophylls, β-carotene, reference blank.
3. Cut out each of the five bands on the chromatogram and place in the appropriate cuvette; place the two xanthophyll bands, violaxanthin and lutein, in the same cuvette. Thus, you will be eluting four pigments or pigment groups: chl *b*, chl *a*, xanthophylls, and β-carotene. You will have to cut the bands into thin strips in order for them to fit in the cuvette.
4. Add 4.0 ml acetone* to each cuvette and cork each tube. Allow the pigments to elute for 5 min, occasionally swirling the tubes. Invert the tubes twice to mix the contents; then remove the paper strips with a pair of forceps, draining the paper against the side of the tube.
5. Measure the absorbance of each sample at the wavelengths listed on Data Sheet 10.2. The spectrophotometer must be adjusted for the reference blank prior to reading the absorbance *at each wavelength*, but all four cuvettes can be read with the one adjustment for the blank at that wavelength. Record the absorbance readings on Data Sheet 10.2

REFERENCES

Goodwin, T. W. 1980. *The Biochemistry of the Carotenoids*, 2nd ed., Vol. I. Chapman and Hall, London.

Govindjee and Govindjee, R. 1974 (Dec.). The absorption of light in photosynthesis. *Sci. Am.* 231:68–82.

Gregory, R. P. F. 1977. *Biochemistry of Photosynthesis*, 2nd ed. John Wiley, New York.

Sherma, J. and Zweig, G. 1971. *Paper Chromatography and Electrophoresis*, Vol. II. Academic Press, New York.

CAUTION: Acetone is toxic; do not pipet by mouth.

EXERCISES AND QUESTIONS

1. Using the distance markings recorded on Data Sheet 10.1, calculate the R_f value for each pigment. Record the distances and the R_f values in the table on Data Sheet 10.1

2. What do the R_f values indicate about the relative solubilities of the pigments in the water and solvent phases?

   _____

   _____

   _____

   _____

3. Explain the relative solubilities of chl b and chl a in the water and solvent phases on the basis of molecular structure.

   _____

   _____

   _____

   _____

4. Explain the relative solubilities of the three carotenoids in the water and solvent phases on the basis of molecular structure.

   _____

   _____

   _____

   _____

   _____

5. On Graph 10.1, plot absorbance (ordinate) versus wavelength (abscissa) for chl b and chl a. Turn the graph paper sideways so that the ordinate is the short axis of the graph paper and the abscissa is the long axis of the graph paper. Each curve should be carefully drawn with a French curve template and then labeled.

6. On Graph 10.2, plot absorbance (ordinate, short axis of graph paper) versus wavelength (abscissa, long axis of graph paper) for the xanthophylls and for β-carotene. Each curve should be carefully drawn with a French curve template and then labeled.

7. At what wavelengths is absorption a maximum for the various pigments?

   chl b _____; chl a _____; xanthophylls _____; β-carotene _____

8. From the absorbance of chl *a* at 663 nm and of chl *b* at 645 nm and the Beer–Lambert equation, determine the concentration of chl *a* and chl *b* in the samples.

[chl *a*] = _____ mg/ml; [chl *b*] = _____ mg/ml. Show the calculations.

9. What is the ratio [chl *a*]/[chl *b*] in spinach? _____

CHROMATOGRAPHY OF PHOTOSYNTHETIC PIGMENTS

Data Sheet 10.1
R_f values for spinach pigments

Pigment	Distance (mm) from origin to solute band	Distance (mm) from origin to solvent front	R_f
β-Carotene			
Lutein			
Violaxanthin			
Chl a			
Chl b			

Data Sheet 10.2
Absorbance readings for chl *b*, chl *a*, xanthophylls, and β-carotene.

Wavelength (nm)	Chl *b*	Chl *a*	Xanthophylls	β-Carotene
400				
410				
420				
430				
440				
450				
460				
470				
480				
490				
500				
520				
540				
560				
580				
600				
620				
640				
645				
660				
663				
680				
700				

Graph 10.1

Absorption spectra for chl *b* and chl *a*.

Graph 10.2

Absorption spectra for the xanthophylls and β-carotene.

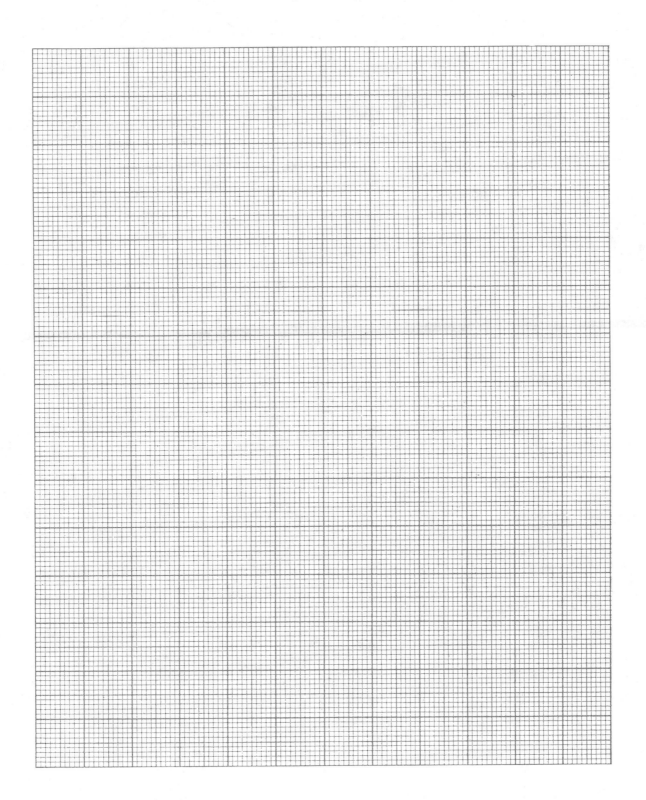

THE HILL REACTION IN ISOLATED CHLOROPLASTS

INTRODUCTION

Photosynthesis is the process by which plants convert light energy to chemical energy. The familiar equation

$$6CO_2 \; + \; 12H_2O \xrightarrow[\text{chloroplasts}]{\text{light}} C_6H_{12}O_6 \; + \; 6O_2 \; + \; 6H_2O$$

summarizes the process but gives little indication of its complexity. There are actually two sets of reactions, the *light reactions* and the *dark reactions*. As their names imply, one reaction series requires light, while the other does not. The products of the light reactions are molecular oxygen, ATP, and reduced electron carrier. The ATP and reduced carrier are used in the dark reactions to convert carbon dioxide to carbohydrates. This project examines the *Hill reaction*, an important phase of the light reactions. The Hill reaction is the transfer of electrons from water to an electron acceptor in the presence of light and chloroplasts. You will follow the rate of the Hill reaction under several experimental conditions.

LIGHT REACTIONS

All components of the light reactions are located in the thylakoid membranes of the chloroplast (Fig. 8.4, p. 91), while the enzymes of the dark reactions are located in the surrounding stroma. The light reactions of photosynthesis are summarized in Figure 11.1. The figure depicts the flow of electrons in the light reactions, with the height of the various compounds indicating the relative energy level of electrons at each point in the pathway. At the heart of the reaction series are two *photosystems*, designated *photosystem II* (*PS II*) and *photosystem I* (*PS I*). Each photosystem is a light-absorbing assembly of chlorophylls *a* and *b*, carotenoids, cytochromes, and other electron carriers. Each photosystem also contains a special chlorophyll *a* molecule that traps the energy absorbed by the other pigment molecules. This chlorophyll *a* molecule

FIGURE 11.1
Path of electrons in the
light reactions of
photosynthesis. The heavy
arrows represent electron
transport chains.

absorbs light at a longer wavelength and, hence, at a lower energy level than do the other chlorophyll molecules. In PS II, the energy trap is a chlorophyll *a* molecule that absorbs maximally at 680 nm. In PS I, it is a chlorophyll *a* molecule that absorbs maximally at 700 nm. These special pigment molecules are designated *P680* and *P700*, respectively. The excited electrons in P680 of PS II are transferred to a *plastoquinone* (PQ in Fig. 11.1), and those in P700 of PS I are transferred to *ferredoxin* (Fd in Fig. 11.1), an iron–sulfur protein.

Linking the two photosystems is a major electron transport chain. As the cytochromes and other electron carriers of this chain are successively reduced and oxidized, some of the energy in the electrons is coupled to the synthesis of ATP. The electrons ejected from P700 (in PS I) pass through another electron transport chain and are ultimately passed to $NADP^+$, the oxidized form of *nicotinamide adenine dinucleotide phosphate*. There is yet one other transfer of electrons, namely, from water to the oxidized P680 of PS II. In a process termed *photolysis*, the details of which are poorly understood, water is split into oxygen, protons, and electrons. These electrons pass via a short electron transport chain to P680.

The flow of electrons from water to $NADP^+$ is referred to as the *noncyclic pathway* since the electrons start in one compound (water) and end in another (NADP). The synthesis of ATP via this route is termed *noncyclic photophosphorylation*. The products of the noncyclic pathway of the light reactions are molecular oxygen, reduced electron carrier (NADPH + H^+), and ATP. ATP is also produced when electrons take a *cyclic pathway* (Fig. 11.1), which involves PS I and electron carriers in the major electron transport chain. It is a true cyclic flow since the electrons emanate from and return to the same compound: P700. The synthesis of ATP via the cyclic pathway is termed *cyclic photophosphorylation*.

The Hill Reaction.   In 1937, Robert Hill showed that isolated chloroplasts can evolve oxygen in the absence of CO_2. This finding was one of the first indications that the source of the electrons in the light reactions is water. In his *in vitro* system, Hill provided an *artificial* electron acceptor. The artificial acceptor intercepts the electrons before they reach PS I, but after the major electron transport chain. The path of electrons from water to the artificial acceptor (A) is, thus,

Various dyes can be used as the artificial electron acceptor so that a general equation, known as the *Hill reaction*, can be written

$$H_2O \ + \ A \xrightarrow[\text{chloroplasts}]{\text{light}} AH_2 \ + \ \tfrac{1}{2}O_2$$

The Hill reaction is formally defined as the photoreduction of an electron acceptor by the hydrogens of water, with the evolution of oxygen. *In vivo*, the final electron acceptor in the light reactions is $NADP^+$.

In isolated chloroplasts, a convenient method for measuring the rate of the Hill reaction is to use as the artificial electron acceptor a dye that changes color as it is reduced. We will use the same dye used in Project 9, 2,6-dichlorophenolindophenol (DCIP), which is blue in its oxidized form and colorless in its reduced form. The change in absorbance, which is measured at 600 nm, will be used to measure the rate of the Hill reaction under a variety of conditions in two experiments. The change in absorbance will be measured at 1-min intervals of exposure to an intense light source. Since the DCIP will begin to revert to its oxidized (blue) state as soon as the chloroplasts (in the reaction vessel) are removed from the light path, it is essential that all absorbance readings be taken as quickly as possible.

EXPERIMENT 1: EFFECT OF INHIBITORS ON THE HILL REACTION

In the first experiment, you will measure the normal rate of the Hill reaction and compare it to the rates in the presence of two inhibitors. One inhibitor is *ammonia*, which is an *uncoupler*, a compound that separates the process of phosphorylation from electron transport. The electron transport system functions, but no ATP is produced. Uncoupling electron transport from ATP synthesis causes an increased electron flow. The dye DCIP can itself be an effective uncoupler. However, at the low concentration used in the present study, there is still *partial* coupling so that the effects of other uncouplers can still be measured. The second inhibitor you will study is *3-(3,4-dichlorophenyl)-1,1-dimethylurea (DCMU)*, an herbicide. DCMU blocks both electron transport *and* phosphorylation by interrupting electron flow at the beginning of the major electron transport chain. The control in Experiment 1 is a reaction mixture kept in the dark.

EXPERIMENT 2: EFFECT OF LIGHT INTENSITY ON THE HILL REACTION

In the second experiment, which you will design yourself, you will determine the effect of light intensity on the rate of the Hill reaction. You will vary the light intensity by placing the reaction vessel at different distances from the light source. Keep in mind that light intensity decreases as the square of the distance from the source.

ISOLATION OF CHLOROPLASTS

The chloroplasts will be obtained from spinach leaves using a modification of a standard fractionation procedure (Whatley and Arnon, 1962), as described in Project 8. You will homogenize the spinach in a mortar with a pestle (Fig. 8.1, p. 88) with a buffered, isotonic salt solution and a small quantity of sand. After squeezing the homogenate through cheesecloth to remove the larger pieces, you will centrifuge the filtrate at 200 g for 1 min, sedimenting debris and whole cells. You will then centrifuge the supernatant at 1300 g for 5 min, sedimenting most of the chloroplasts (and nuclei). For all cell fractionation procedures, the solutions and containers must be ice-cold.

PROCEDURES

For all procedures, you should work in pairs.

ISOLATION OF THE CHLOROPLAST FRACTION

1. Weigh out 4 g of fresh spinach leaves from which the major veins have been removed.
2. Cut the leaves into small pieces with scissors and place in a chilled mortar with 15 ml of ice-cold *NaCl-buffer* and a sprinkling of purified sand. Grind the tissue with a chilled pestle for 2 min.
3. Filter the suspension through four layers of cheesecloth into a chilled 15-ml conical centrifuge tube. Also wring out the juice into the centrifuge tube.
4. Centrifuge the filtrate at 200 g for 1 min. Make sure that the centrifuge tubes are balanced; the tubes opposite each other should have the same total volume.
5. Decant the supernatant into a clean, chilled centrifuge tube and spin at 1300 g for 5 min. Remember to balance the tubes.
6. Decant and discard the supernatant and then, using a graduated cylinder, add 10 ml of ice-cold NaCl-buffer to the pellet in the centrifuge tube. With a Pasteur pipet, thoroughly resuspend the pellet. To ensure that the chloroplast suspension is thoroughly mixed, cover the centrifuge tube with Parafilm and invert several times.
7. Transfer 4.0 ml of the chloroplast suspension* to a clean, chilled test tube and dilute with 6.0 ml of ice-cold NaCl-buffer*. Cover the test tube, invert several times, and place in an ice-water bath, in which it should remain during the entire experiment. It is the *diluted* chloroplast suspension (hereafter referred to as *chloroplast suspension*) that will be used to measure the Hill reaction. The *original* (concentrated) chloroplast suspension should be saved (in an ice-water bath) in the event that the Hill reaction proceeds too slowly using the 4:10 dilution.

*CAUTION: *Do not pipet by mouth.*

EXPERIMENT 1: EFFECT OF INHIBITORS ON THE HILL REACTION

1. Allow the spectrophotometer to warm up for at least 5 min. Instructions for its use are given in Project 4, pages 30 and 35 (steps 8 and 9 at the bottom). Set the wavelength at 600 nm.

2. Prepare a water bath as follows. Add 150 ml of water at 20° C to a 250-ml beaker. During illumination, the tubes (cuvettes) are kept in the water bath to keep the reaction temperature fairly constant. Adjust the temperature of the water bath at the beginning of each experimental trial.

3. Place the water bath 25 cm from a lamp with a 100-watt frosted incandescent bulb, but do not turn on the lamp until the start of each experimental trial.

4. Label five cuvettes as shown in the following table. Except for the ice-cold chloroplast suspension, all solutions should be at room temperature.

Tube	NaCl-buffer*	DCIP* (4×10^{-4} M)	Ammonia* (0.01 N)	DCMU* (10^{-4} M)	Distilled water	Chloroplast suspension*
Blank	3.5 ml	—	—	—	1.0 ml	0.5 ml
1[†]	3.5 ml	0.5 ml	—	—	0.5 ml	0.5 ml
2	3.5 ml	0.5 ml	—	—	0.5 ml	0.5 ml
3	3.5 ml	0.5 ml	0.5 ml	—	—	0.5 ml
4	3.5 ml	0.5 ml	—	0.5 ml	—	0.5 ml

*CAUTION: *Do not pipet by mouth.*
[†]Tube 1 will be wrapped with two layers of aluminum foil so that no light can enter.

5. Wrap tube 1, the nonilluminated control, with two layers of aluminum foil and make a loose-fitting foil cap for the top of the cuvette.

6. Prepare the reference blank and tube 1. Add the solutions* in the sequence given across the top of the table, from left to right. *Before and after* adding the chloroplast suspension, which should be inverted several times just before it is added, cover each cuvette with Parafilm and invert twice to mix the contents. Remove the Parafilm and cover tube 1 with the foil cap. Note the time and set tube 1 aside. An absorbance reading should be taken after 10 min but, meanwhile, proceed with the steps that follow.

7. Adjust the spectrophotometer for the blank.

8. Prepare tube 2. Add the solutions in the same manner used for tube 1 in step 6. Immediately take the 0-min absorbance reading and enter on Data Sheet 11.1.

9. Immediately place tube 2 in the water bath, turn on the lamp, and note the time. After 1 min illumination, remove the tube from the water bath, and *quickly* wipe the surface of the cuvette and measure the absorbance. All absorbance readings must be made as quickly as possible. Enter the absorbance reading on Data Sheet 11.1

10. Return tube 2 to the water bath and take absorbance readings at 1-min intervals of illumination for 10 min. Enter the absorbance readings on Data Sheet 11.1. The spectrophotometer should be adjusted for the blank after every few readings. Remember to take the 10-min absorbance reading for tube 1. Enter it on Data Sheet 11.1. The absorbance should reach zero (or close to it) in 7–10 min.

11. Prepare tube 3, which contains the uncoupler (ammonia). Add the solutions in the same manner used for tube 1 in step 6. Take an absorbance reading immediately and then at 1-min intervals of illumination for 10 min. Remember

*CAUTION: *Do not pipet by mouth.*

to adjust the temperature of the water bath to 20° C at the outset and to adjust the spectrophotometer for the blank after every few readings. Enter the absorbance readings on Data Sheet 11.1

12. Prepare tube 4, which contains DCMU. Add the solutions in the same manner used for tube 1 in step 6. Take an absorbance reading immediately and then at 1-min intervals of illumination for 10 min. Remember to adjust the temperature of the water bath to 20° C at the outset and to adjust the spectrophotometer for the blank after every few readings. Enter the absorbance readings on Data Sheet 11.1

EXPERIMENT 2: EFFECT OF LIGHT INTENSITY ON THE HILL REACTION

Design and carry out an experiment to measure the normal rate of the Hill reaction at different light intensities. In the space provided, describe each of the tubes in your experiment, including the contents of each and the distance from the light source. Enter all absorbance readings on Data Sheet 11.1.

_____

_____

_____

_____

_____

_____

REFERENCES

Gregory, R. P. F. 1977. *Biochemistry of Photosynthesis*, 2nd ed. John Wiley, New York.

Izawa, S. and Good, N. E. 1972. Inhibition of photosynthetic electron transport and photophosphorylation. In *Methods in Enzymology*, Vol. 24, San Pietro, A., ed., pp. 355–377. Academic Press, New York.

Trebst, A. 1972. Measurement of the Hill reaction and photoreduction. In *Methods in Enzymology*, Vol. 24, San Pietro. A., ed., pp. 146–165. Academic Press, New York.

Whatley, F. R. and Arnon, D. I. 1962. Photosynthetic phosphorylation in plants. In *Methods in Enzymology*, Vol. 6, Colowick, S. P. and Kaplan, N. O., eds., pp. 308–313. Academic Press, New York.

EXERCISES AND QUESTIONS

1. On Data Sheet 11.2, enter the *total change in absorbance* (ΔA) at each time interval, i.e., the difference between the initial absorbance reading in each tube and the absorbance reading at the specified time.

2. On Graph 11.1, plot the rate of the Hill reaction for tubes 2–4 (Experiment 1). Plot the total change in absorbance (ordinate) versus time (abscissa). Include the origin as a point for all curves, i.e., the 0-min value for ΔA is assumed to be 0. Draw the best-fit curve with a ruler and French curve template and label each plot.

3. In which tube in Experiment 1 does the reaction proceed most rapidly?

   _____ Explain. _____

   _____

   _____

4. Explain the appearance of the curve for the tube with the DCMU (tube 4).

   _____

   _____

   _____

5. Is there any evidence that the Hill reaction proceeded in tubes 2 and 3 prior to the

   time that the 0-min reading was taken? _____

   Explain. _____

   _____

   _____

6. On Graph 11.2, plot the rate of the Hill reaction for the tubes in Experiment 2. Label each plot with the distance from the reaction vessel to the light source.

7. Discuss the relationship between the rate of the Hill reaction and light intensity. Include in your discussion any possible sources of error in Experiment 2.

   _____

   _____

   _____

   _____

   _____

Data Sheet 11.1

Absorbance Readings (A_{600}) at each time interval.

EXPERIMENT 1

Time (min)

Tube	0	1	2	3	4	5	6	7	8	9	10
1	—	—	—	—	—	—	—	—	—	—	
2											
3											
4											

EXPERIMENT 2

Time (min)

Tube	0	1	2	3	4	5	6	7	8	9	10

DESCRIPTION OF TUBES AND EXPERIMENTAL CONDITIONS IN EXPERIMENT 2

Data Sheet 11.2

Total change in absorbance (ΔA) at each time interval.

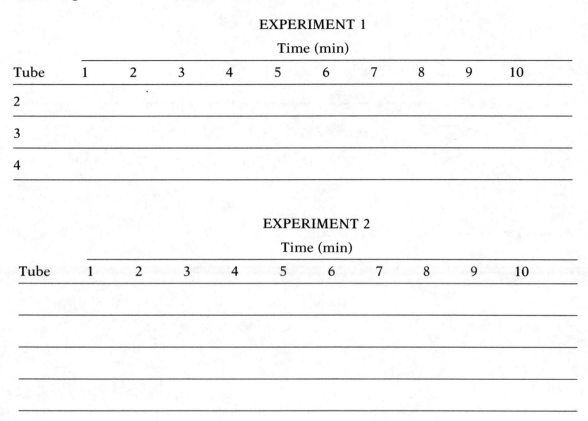

EXPERIMENT 1

Time (min)

Tube	1	2	3	4	5	6	7	8	9	10
2										
3										
4										

EXPERIMENT 2

Time (min)

Tube	1	2	3	4	5	6	7	8	9	10

Graph 11.1
Plot of total change in absorbance (ΔA) versus time (min) for tubes 2–4 (Experiment 1).

Graph 11.2
Plot of total change in absorbance (ΔA) versus time (min) for the tubes in Experiment 2.

CELL MOTILITY

INTRODUCTION

It is movement, perhaps more than any other attribute, that distinguishes living from nonliving systems. The locomotion can take many forms. Flagella are whiplike cell appendages that enable spermatozoa and many protozoans to swim. Other protozoans are propelled through their aqueous environment by the coordinated beating of cilia, which are usually shorter than flagella and present in greater numbers on an individual cell. A slower form of cell motility is amoeboid movement, which occurs by a flow of cytoplasm into extensions of the cell surface. It is observed not only in amoebae but in many metazoan cell types, such as leukocytes. Protozoa are ideal subjects for observing and investigating the foregoing modes of locomotion as they are large, easily handled, and readily observed in the living state. In this project, you will observe amoeboid movement in an amoeba (*Amoeba proteus* or *Pelomyxa carolinensis*), ciliary motion in *Paramecium caudatum*, and flagellar motion in *Chlamydomonas reinhardi*. In addition, you will perform an experiment with *Chlamydomonas*, relating flagellar length to the ability of the organism to swim.

AMOEBOID MOVEMENT

The term *amoeboid movement* refers to locomotion resulting from the streaming of cytoplasm into advancing extensions of the cell surface, the *pseudopods*. Amoeboid movement is exhibited not only by amoebae but by various metazoan cells. For example, certain leukocytes move through body tissues and engulf bacteria, and embryonic cells migrate to new regions of the embryo during morphogenesis. This mode of locomotion is most readily observed and studied in amoebae; they are very large (up to several millimeters!), and cytoplasmic changes are obvious with ordinary bright-field or phase-contrast optics.

The familiar freshwater species *Amoeba proteus* (Fig. 12.1) and the larger *Pelomyxa carolinensis* have pseudopods with round ends. The thin peripheral cytoplasm, or *ectoplasm*, is almost devoid of granules and vacuoles. It also has a rather viscous or gelated consistency (*plasmagel*). The internal cytoplasm, or *endoplasm*, is very granular and contains the major organelles. The endoplasm is quite fluid; it is in a solated state (*plasmasol*).

Amoeboid locomotion is accompanied by striking sol–gel transformations at the anterior and posterior ends of the cell (Fig. 12.1). At the anterior end, just behind a thick region of clear cytoplasm termed the *hyaline cap*, the advancing endoplasm spreads laterally and posteriorly, much like the flow of water at the apex of a fountain. As the endoplasm turns posteriorly to become part of the ectoplasm, there is a conversion from plasmasol to plasmagel. At the posterior end, or *uroid*, there is the reverse conversion, from plasmagel to plasmasol.

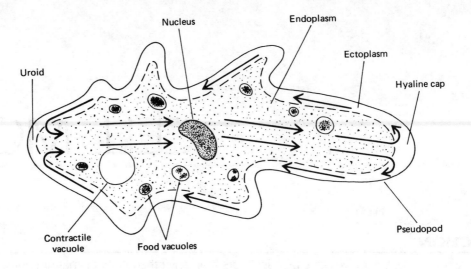

FIGURE 12.1
Amoeba proteus, length approximately 500 μm. Arrows indicate the direction of cytoplasmic flow.

Labels in figure: Nucleus, Endoplasm, Ectoplasm, Uroid, Hyaline cap, Contractile vacuole, Food vacuoles, Pseudopod

While the mechanism responsible for amoeboid movement is poorly understood, it now appears that *actin microfilaments* and possibly *myosin*, as well, are involved in generating the force. Both types of molecules have been extracted from amoebae, and, in electron micrographs, actin-containing microfilaments are visible in the plasmagel. Relating the meshwork of observed microfilaments to amoeboid movement has been perplexing since there is no evidence of an orderly array of filaments, as occurs in striated muscle.

Several models have been proposed to explain amoeboid movement. Two models that address the basic issue of where the contractile force is generated will be examined in this project. In the *tail contraction model*, proposed by S. O. Mast in 1926, the contractile force is generated by the gelated ectoplasm in the uroid region. According to this model, the endoplasm is *pushed* forward into the advancing pseudopod. In 1961, R. D. Allen proposed the *frontal contraction model*, in which the endoplasm is *pulled* forward by a contractile force generated at the tip of the advancing pseudopod. In the first set of observations in this project, you will examine the pattern of cytoplasmic flow in various regions of an amoeba and deduce which of these two models your observations support.

For studying live amoebae, which are rather thick specimens, you will need to use a *Vaseline chamber*. The coverslip in this kind of wet mount is elevated and sealed by a ledge of petroleum jelly placed around its edges. The amoebae will have room to move about and can be observed over an extended period of time without drying out.

CILIARY MOTION AND FLAGELLAR MOTION

There is a basic ultrastructure that is common to both cilia and flagella (Fig. 12.2). The individual cilium or flagellum contains *microtubules*, which are unbranched, hollow cylinders, about 24 nm in diameter, composed entirely of protein. The microtubules are always observed in a characteristic array of nine outer doublets or pairs of microtubules and two central single microtubules. The "9 + 2" core is referred to as the *axoneme*. The axoneme is enclosed by an extension of the plasma membrane so that each cilium and flagellum is really a protruding region of the cytoplasm. At the base of each cilium and flagellum, there is a *basal body*, a barrel-shaped organelle that has the same basic ultrastructure as a centriole. There is compelling evidence that ciliary and flagellar motion results from the doublets sliding past each other (Satir, 1974). Electron micrographs reveal that the doublets on the inner and outer curvature of the bent cilium are the same length but that the doublets on the inner curvature are always closer to the tip of the cilium (Fig. 12.2).

A general distinction between cilia and flagella is that the former are short (5–10 μm) and numerous, whereas the latter are long (up to 150 μm) and sparse. Another

154

"9 + 2" core of
microtubules

Plasma
membrane

FIGURE 12.2
Organization of
microtubules in cilia and
flagella of eukaryotes.

difference is the nature of their motion. Cilia exhibit an oarlike beat, with alternating power and recovery strokes (Fig. 12.3a). Flagella, on the other hand, usually undulate, with successive waves moving from the base of the flagellum to the tip (Fig. 12.3b). This wave form is observed in the flagella of swimming spermatozoa as well as in many protozoans. Flagella can also exhibit other patterns of movement, as you will observe in *Chlamydomonas*.

Observations on Cilia in *Paramecium*.   Ciliary beating will be studied in *Paramecium* (Fig. 12.4). It is probably the most familiar of the ciliates, with its characteristic cigar shape and cilia-covered surface. You will study its swimming motions and the process by which fluid and particles are swept over the surface and into the *buccal cavity* (Fig. 12.4) as the organism feeds. The latter process is similar to the action of the ciliated columnar epithelial cells that line the bronchi and trachea; the beating cilia of these epithelial cells aid in removal of mucous and suspended foreign particles from the respiratory passages.

To make microscopic observations of paramecia, you will have to slow them down by adding yeast cells to the wet mount. The paramecia swim rapidly at first but after 10 minutes congregate around air bubbles and yeast, slowing down as they feed. The yeast cells will have previously been stained with *Congo red* to make them clearly visible.

Power stroke

(a)

(b)

Recovery stroke

FIGURE 12.3
Movement of a cilium (a)
and a flagellum (b). The
large arrow indicates the
direction in which the
organism moves.

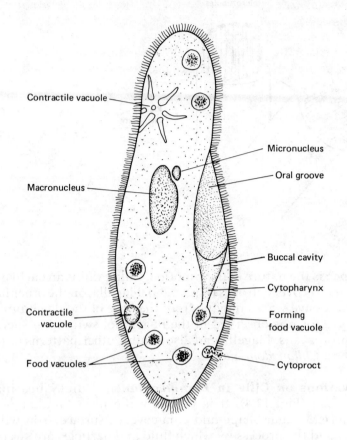

FIGURE 12.4
Paramecium caudatum,
length 170–290 μm.

Contractile vacuole

Micronucleus

Oral groove

Macronucleus

Buccal cavity

Cytopharynx

Contractile
vacuole

Forming
food vacuole

Food vacuoles

Cytoproct

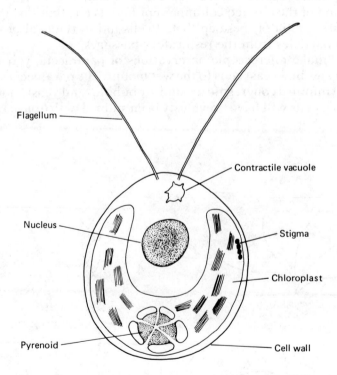

FIGURE 12.5
Chlamydomonas reinhardi,
diameter approximately 10
μm.

Flagellum

Contractile vacuole

Nucleus

Stigma

Chloroplast

Pyrenoid

Cell wall

CELL MOTILITY

Observations on Flagella in *Chlamydomonas*. Flagellar motion will be studied in *Chlamydomonas* (Fig. 12.5). *Chlamydomonas* is usually classified as a green alga for it is photosynthetic with a single, large (about 40% of the cell volume), cup-shaped chloroplast and a thin cell wall. At the anterior end of the organism there are two prominent flagella, which function in locomotion. You will study the motility of *Chlamydomonas* in an ordinary wet mount and in a preparation to which a small drop of *Protoslo* (Carolina Biological Supply Co.) has been added. Protoslo is a viscous liquid that is helpful in slowing the organisms for microscopic examination; however, it may alter the normal flagellar motion.

EXPERIMENT ON FLAGELLAR FUNCTION

Chlamydomonas is an excellent subject for studying the relationship between flagellar function and growth. Its suitability for such studies stems from the fact that *Chlamydomonas* can lose or regenerate flagella, depending on culture conditions and experimental manipulations. For example, when grown on solid medium, such as an agar slant, *Chlamydomonas* lose their flagella and grow as a nonmotile form. When returned to liquid medium, the organisms regenerate new flagella and soon become motile. It is this return of motility, which accompanies the flagellar regeneration, that is the focus of the present experiment. Specifically, you will determine the minimum length that the flagella must be for the organisms to swim.

For the experiment on flagellar function, you will be working with normal, flagellated *Chlamydomonas* that have been growing in liquid medium for several days. The first step in the experiment is the removal of flagella, termed *deflagellation*. Deflagellation is accomplished by subjecting the organisms to *pH shock*, a rapid lowering of the pH of the medium. The flagella fall off into the medium, and the deflagellated cells can be recovered by centrifugation. The deflagellated cells are then transferred to fresh liquid medium in which they soon begin regenerating flagella. Accompanying the regeneration is a return of motility, which you will monitor by scoring the percentage of motile organisms in a wet mount. You will also score the percentage of motile organisms in the original (*nondeflagellated*) culture as a control. When normal motility has returned to the deflagellated culture, i.e., when the percentage of motile cells is approximately the same as that in the control culture, you will fix a sample and determine flagellar length at that time.

PROCEDURES

MICROSCOPIC EXAMINATION OF AMOEBAE

1. Using a dissecting microscope, locate the amoebae in the culture dish. With a dropper, carefully transfer one or two organisms to a clean slide. Examine the slide with the dissecting microscope to be sure you have transferred an amoeba and then add a coverslip that has a ledge of petroleum jelly on its edges for support. A neat way of applying the Vaseline is to rub some on the thumb and index finger and then scrape off a small amount with each edge of the coverslip.

2. With the low power objective of a compound microscope, observe the movement of an amoeba over a 10-min period, sketching outlines of the specimen at 2-min intervals. Use the same size scale in each sketch, which should be made on the drawing paper provided (p. 169). Note the changes in the shape and size of the pseudopods. Using an arrow on each drawing, indicate the overall direction of movement of the entire organism.

3. Observe the cytoplasmic streaming in different regions of a pseudopod with low power and high-dry. Continue observing until the organism changes direction.

Closely examine and compare the direction of cytoplasmic streaming in the old and new pseudopods and then answer the questions on page 161.

MICROSCOPIC EXAMINATION OF *PARAMECIUM*

1. On a clean slide, place a drop of the *Paramecium* culture. Dip the tip of a toothpick in the Congo red-stained yeast suspension and then stir the drop with the toothpick. Only a small amount of the yeast suspension should be added; the drop should turn pink, not red. Add a coverslip and carefully seal the edges with Vaseline applied with a clean toothpick.

2. Scan the preparation under low power. While you are waiting for the paramecia to slow down, observe their normal swimming pattern. Identify the anterior and posterior ends. Note that the organism rotates on its own axis. Under high-dry, examine the ciliary motion in various regions of a slow-moving organism and then answer the question on page 161.

3. Examine the yeast cells near the buccal cavity of a slow-moving or stationary specimen. Then answer the question on page 161.

4. Observe the collection of yeast cells in a *food vacuole* at the blind end of the *cytopharynx* (Fig. 12.4). If you continue to observe the food vacuole, you will be able to follow the course of intracellular digestion, specifically, the changes in the size and pH of food vacuoles. Congo red is a pH indicator that is bright orange-red at pH 5 and above, purplish between pH 5 and pH 3, and blue below pH 3.

MICROSCOPIC EXAMINATION OF *CHLAMYDOMONAS*

1. On a clean slide, place a drop of the *Chlamydomonas* culture and a coverslip. Under low power, observe the normal motion of the swimming organisms.

2. Make a fresh wet mount with a drop of the *Chlamydomonas* culture and a tiny drop of Protoslo. Mix with a toothpick and add a coverslip.

3. Examine under low power and high-dry. Close the condenser iris until the flagella are visible. Examine some slowed organisms that are moving forward, i.e., with the flagella at the leading end. Then, on page 161, describe the motion of the flagella.

EXPERIMENT ON FLAGELLAR FUNCTION

Except for the deflagellation, you should work individually. The deflagellation procedure should be carried out by one designated person, since a single deflagellated culture provides enough material for the entire class.

Setting up the Tubes.

1. Place four test tubes in your test tube rack and label them as follows: *Deflagellated/sample 1, Nondeflagellated/sample 1, Deflagellated/sample 2, Nondeflagellated/sample 2.*

2. To each tube, add one drop of Lugol's iodine solution (a fixative and stain) and three drops of culture medium (Medium I of Sager and Granick). These tubes will be used to fix the samples at the outset and at the time when motility has returned.

Deflagellation Procedure.

1. From an actively growing *Chlamydomonas* culture, transfer a 40-ml aliquot to a 100-ml beaker containing a magnetic stirrer. Save the remainder of the nondeflagellated culture as a control.

2. Deflagellate the organisms in the beaker by rapidly lowering the pH of the medium. Monitor the pH with the electrodes of a pH meter. With constant stirring, lower the pH to 4.5 within 30 sec by adding dropwise 0.5 N acetic acid.

3. Wait 30 sec and then restore the pH of the culture to 6.8 by adding dropwise 0.5 N KOH.

4. Pour 10 ml of the deflagellated culture into a 15-ml graduated, conical centrifuge tube. Add 10 ml of water to another centrifuge tube to balance the head and then centrifuge at 1300 *g* for 5 min.

5. Decant and discard the supernatant liquid, which contains the flagella. Using a graduated cylinder, add 10 ml of Medium I to the pellet, which contains the deflagellated cells. Resuspend the cells with a Pasteur pipet.

6. Pour the contents of the centrifuge tube into a 50-ml Erlenmyer flask labeled *Deflagellated*. Also pour a 10-ml aliquot from the original culture into another 50-ml flask labeled *Nondeflagellated*. Place the two cotton-plugged flasks in a gently reciprocating shaker that is illuminated with a fluorescent lamp and proceed immediately to step 1, following.

Determination of the Return of Motility.

1. As soon as possible, remove one or two drops from the deflagellated culture with a Pasteur pipet, add to the test tube labeled *Deflagellated/sample 1*, and swirl the tube to fix the cells rapidly. Using a fresh Pasteur pipet, remove one or two drops from the nondeflagellated culture, add to the test tube labeled *Nondeflagellated/sample 1*, and swirl the tube. Record the time on Data Sheet 12.2. Put these samples aside until the last two samples have been taken. Then all the samples will be examined for flagellar length.

2. Prepare two Vaseline chambers as follows. Place two drops of the deflagellated and nondeflagellated cultures on separate, clean slides, and to each add a coverslip with a ledge of petroleum jelly.

3. Using the low power objective, examine the "Nondeflagellated" Vaseline chamber for the percentage of motile cells, i.e., the percentage of organisms that are swimming in the microscope field. Count the number of motile and nonmotile cells in three microscope fields, recording the total number of each and the percentage of motile cells for this initial scoring on Data Sheet 12.1. Do the same scoring for the "Deflagellated" Vaseline chamber.

4. Repeat the scoring every 5 min until the percentage of motile cells in the "Deflagellated" Vaseline chamber is approximately the same as that in the "Nondeflagellated" chamber. Immediately after normal motility has returned to the deflagellated cells, fix *sample 2* from the deflagellated and nondeflagellated cultures, using the appropriately labeled test tubes and fresh Pasteur pipets. Record the time on Data Sheet 12.2.

Measurement of Flagellar Length.

1. With a fresh Pasteur pipet, gently resuspend the fixed sample and place a drop on a clean slide. Add a coverslip and examine under low power to locate an area with a concentration of cells.

2. Using a calibrated ocular micrometer with the oil-immersion objective, measure flagellar length in each of 20 cells. Select only cells that have at least one fairly straight flagellum. Since both flagella will be about the same length, just

measure the straighter of the two. Record the measurements for the four samples on Data Sheet 12.2

3. Proceed to Exercises and Questions, page 163.

REFERENCES

Allen, R. D. 1981. Motility. *J. Cell Biol. 91*:148s–155s.

Allen, R. D. and Taylor, D. L. 1975. The molecular basis of amoeboid movement. In *Molecules and Cell Movement*, Inoué, S. and Stephens, R. E., eds., pp. 239–258. Raven Press, New York.

Gibbons, I. R. 1981. Cilia and flagella of eukaryotes. *J. Cell Biol. 91*:107s–124s.

Satir, P. 1974 (Oct.). How cilia move. *Sci. Am. 231*:45–52.

OBSERVATIONS AND QUESTIONS

MICROSCOPIC EXAMINATION OF AMOEBAE

When a change in the direction of cytoplasmic streaming occurs, is the change first apparent in the *old* or *new* pseudopod?

_____

In what part of the pseudopod is the movement first observed?

_____

Which of the two hypotheses—tail contraction model or frontal contraction model—do your observations support?

_____

Explain.

_____

_____

_____

_____

MICROSCOPIC EXAMINATION OF *PARAMECIUM*

Do all groups of cilia beat identically? _____ Explain.

_____

_____

Do all yeast cells that enter the buccal cavity get ingested? _____ Explain.

_____

_____

MICROSCOPIC EXAMINATION OF *CHLAMYDOMONAS*

What swimming stroke does the coordinated movement of both flagella resemble?

_____

EXERCISES AND QUESTIONS

EXPERIMENT ON FLAGELLAR FUNCTION

1. From the time you fixed the first samples, how long did it take for normal

   motility to return to the deflagellated organisms? _____ min Was the return of

   motility sudden or gradual? _____ What does this suggest about
   flagellar function in these organisms?

   _____

   _____

   _____

2. For each set of length measurements, calculate the mean, both in ocular
   micrometer units and in micrometers. Enter the values on Data Sheet 12.2. What
   was the length of the flagella when normal motility returned to the deflagellated

   organisms? _____ μm

3. What percentage of the length in controls must be attained before normal

   motility returns? _____ %

4. For the time period you studied, what is the mean rate of flagellar regeneration in

   Chlamydomonas? _____ μm/min

Data Sheet 12.1
Frequency of motile and nonmotile cells in the deflagellated and nondeflagellated cultures.

Deflagellated

Time (min)	Number of motile cells				Number of nonmotile cells				Total	% Motile cells
Initial										
5										
10										
15										
20										
25										
30										

Nondeflagellated

Time (min)	Number of motile cells				Number of nonmotile cells				Total	% Motile cells
Initial										
5										
10										
15										
20										
25										
30										

Data Sheet 12.2

Flagellar length measurements in the deflagellated and nondeflagellated cultures.

Deflagellated

Sample Time	Flagellar length measurements (ocular micrometer units) in 20 cells																				Mean	Mean (μm)
1																						
2																						

Nondeflagellated

Sample Time	Flagellar length measurements (ocular micrometer units) in 20 cells																				Mean	Mean (μm)
1																						
2																						

ASSEMBLY OF FLAGELLAR MICROTUBULES

INTRODUCTION

Nearly every eukaryotic cell contains microtubules. They are an important component of the cytoskeleton and are the major element in spindle fibers, cilia, and flagella. Each microtubule is a hollow, unbranched cylinder composed of globular protein subunits. Perhaps the most intriguing feature of microtubules is their ability to assemble from the globular protein subunits in the cytoplasmic pool. During microtubule assembly, the subunits add onto the ends of existing microtubules, thereby increasing their length. Flagellar regeneration, as observed in Chlamydomonas *in the preceding project, is basically a process of microtubule assembly. In this project, you will study the kinetics of normal flagellar regeneration in* Chlamydomonas *and the effects produced by two chemicals, colchicine and cycloheximide.*

THE MICROTUBULE AND ITS ASSEMBLY

All microtubules have a similar ultrastructure. In cross section, the wall of each microtubule is composed of 13 globular subunits (Fig. 13.1). The wall thickness is 5 nm, and the total diameter of the microtubule is about 24 nm. Along the length of the microtubule, which can measure 25 μm or more, the globular subunits form 13 rows, each row termed a *protofilament*.

Ultrastructural analyses of the microtubule have shown that the wall consists of pairs of globular subunits, i.e., *dimers*. Biochemical studies have revealed that there are two types of globular subunits in the dimer, so it is appropriately termed a *heterodimer*. Each heterodimer is composed of one molecule of *α-tubulin* and one molecule of *β-tubulin* (Fig. 13.1). Both are proteins with similar molecular weights (~50,000 daltons) and related amino acid sequences. The heterodimers in the microtubule wall are arranged in a characteristic way, with the α- and β-tubulins alternating along the protofilament and slightly staggered in adjacent protofilaments so as to form a helical array (Fig. 13.1).

In 1972, it was discovered by R. C. Weisenberg that tubulin extracted from cells could assemble into microtubules in a test tube. For such *in vitro* assembly to occur, both *guanosine triphosphate (GTP)* and Mg^{2+} must be present. The process of microtubule assembly is greatly facilitated if *microtubule-associated proteins*, or *MAPs*, are present. MAPs are proteins that remain bound to the tubulin after its extraction from cells. Both assembly and disassembly can occur in the same *in vitro* system. Factors that shift the equilibrium toward disassembly include a relatively high concentration of Ca^{2+} and low temperature.

While the precise conditions that favor microtubule assembly or disassembly in the cell are not known for certain, assembly *in vivo* probably involves the addition of tubulin dimers onto one end of the microtubule. This is depicted in Figure 13.1. The

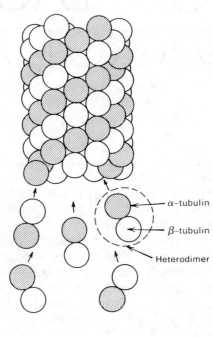

α-tubulin

β-tubulin

Heterodimer

FIGURE 13.1
Diagram showing the arrangement of α-tubulin and β-tubulin in the microtubule and the polymerization of heterodimers at the polymerization end of the microtubule.

figure does not show the process of disassembly, which is believed to occur at the opposite end of the microtubule. Thus, there is a polarity to the microtubule, with polymerization occurring at one end and depolymerization occurring at the opposite end. Flagellar microtubules are relatively stable structures compared, for example, to the microtubules in spindle fibers. In flagella, there appears to be very little depolymerization at the depolymerization end of the microtubules. In cells that are regenerating flagella, it has been shown that the tubulin dimers add on at the *distal* end of the growing flagellum (Rosenbaum, Moulder, and Ringo, 1969).

EXPERIMENT ON FLAGELLAR REGENERATION

The present experiment is based on a study by Rosenbaum, Moulder, and Ringo (1969). It examines the effects of colchicine and cycloheximide on flagellar regeneration in *Chlamydomonas reinhardi*. Colchicine is known to bind to free dimers in the cytoplasmic pool. The colchicine-bound dimers are able to attach to the growing ends of the microtubules but, once attached, prevent any further addition of dimers, whether bound to colchicine or not. Cycloheximide is a potent inhibitor of protein synthesis in eukaryotic cells. In the presence of cycloheximide, there is no longer synthesis of α- or β-tubulin.

The first step in the procedure is the removal of flagella. As in the preceding project, deflagellation is accomplished by rapidly lowering the pH of the *Chlamydomonas* culture. Following the pH shock, the deflagellated cells are divided into three aliquots, each of which is incubated in a different growth medium. One aliquot will be allowed to regenerate flagella in fresh culture medium (Medium I of Sager and Granick). A second aliquot will be placed in Medium I with added colchicine, and the third placed in Medium I with added cycloheximide. From each

ASSEMBLY OF FLAGELLAR MICROTUBULES

culture, you will periodically fix samples in Lugol's iodine solution, which also stains the flagella. You will then measure flagellar length in each sample and determine the rate of flagellar regeneration for the three culture conditions. The deflagellated cells incubated in Medium I will serve as a control for measuring the normal rate of regeneration. In addition, the original, nondeflagellated culture will be sampled at the beginning and end of the experiment for a measure of normal flagellar length.

PROCEDURES

EXPERIMENT ON FLAGELLAR REGENERATION

For this experiment, you should work in teams of three, with each member responsible for sampling one of the cultures: Medium I, colchicine, or cycloheximide. The team member responsible for the colchicine culture should also sample and score the nondeflagellated culture. Thus, each team collects data for a complete experiment. The deflagellation procedure should be carried out by one designated team, since a single deflagellated culture provides enough material for the entire class.

Setting up the Tubes.
1. For your particular culture condition, place 10 test tubes in a rack. Note the culture for which you are responsible and label the tubes with the sampling times: 0, 10, 20, 30, 40, 50, 60, 75, 90, and 105 min. The team member responsible for the colchicine culture should also have two additional tubes for sampling the nondeflagellated culture at the start (0 min) and end (105 min) of the experiment.
2. To each tube, add one drop of Lugol's iodine solution and three drops of Medium I.

Deflagellation Procedure.
1. From an actively growing *Chlamydomonas* culture, transfer a 40-ml aliquot to a 100-ml beaker containing a magnetic stirrer. Save the remainder of the nondeflagellated culture for determination of normal flagellar length.
2. Deflagellate the organisms in the beaker by rapidly lowering the pH of the medium. Monitor the pH with the electrodes of a pH meter. With constant stirring, lower the pH to 4.5 within 30 sec by adding dropwise 0.5 N acetic acid.
3. Wait 30 sec and then restore the pH of the culture to 6.8 by adding dropwise 0.5 N KOH.
4. Pour 30 ml of the deflagellated culture into three 15-ml conical centrifuge tubes, 10 ml per tube, and centrifuge at 1300 g for 5 min.
5. Decant and discard the supernatant liquid, which contains the flagella. To the pellets, which contain the deflagellated cells, add 10 ml of each of the following media, carefully noting which tube gets which medium: Medium I,* Medium I + colchicine (3 mg/ml),† or Medium I + cycloheximide (10 µg/ml).†
6. Resuspend the pellets, using a separate Pasteur pipet for each.
7. Pour the contents of each centrifuge tube into a labeled 50-ml Erlenmyer flask. Also transfer a 10-ml aliquot from the original culture (nondeflagellated cells)* to a labeled 50-ml flask. Place the four cotton-plugged flasks in a gently reciprocating shaker that is illuminated with a fluorescent lamp and proceed immediately to step 1 following.

*CAUTION: *Do not pipet by mouth.*
†CAUTION: *Colchicine and cycloheximide are poisonous; do not pipet by mouth.*

Sampling.
1. As soon as possible, fix the 0-min sample as follows. Using a fresh Pasteur pipet, remove one or two drops from the culture and add to the appropriate test tube. Swirl the tube to fix the cells rapidly. Record the time.
2. Begin measuring flagellar length in the 0-min sample, following the measurement procedure below. Score as many cells as possible between samplings; however, the cells can remain in the fixative and can be scored after all the samples have been taken.
3. Continue sampling every 10 min for the first 60 min and then every 15 min thereafter until the 105-min sample has been fixed. Always be sure that the Pasteur pipet contains no liquid from a previous sampling or use a fresh pipet each time.

Measurement of Flagellar Length.
1. With a fresh Pasteur pipet, gently resuspend the fixed sample and place a drop on a clean slide. Add a coverslip and examine under low power to locate an area with a concentration of cells.
2. Using a calibrated ocular micrometer with the oil-immersion objective, measure flagellar length in each of 20 cells. Select only cells that have at least one fairly straight flagellum. Since both flagella will be about the same length, just measure the straighter of the two. Record the measurements for your culture(s) on Data Sheet 13.1.

REFERENCES

Dustin, P. 1980 (Aug.). Microtubules. *Sci. Am. 243*:66–76.

Haimo, L. T. and Rosenbaum, J. L. 1981. Cilia, flagella, and microtubules. *J. Cell Biol. 91*:125s–130s.

Rosenbaum, J. L., Moulder, L. E., and Ringo, D. L. 1969. Flagellar elongation and shortening in *Chlamydomonas*: The use of cycloheximide and colchicine to study the synthesis and assembly of flagellar proteins. *J. Cell Biol. 41*:600–619.

NAME _____ SECTION _____ DATE _____

LAB PARTNERS _____

EXERCISES AND QUESTIONS

1. For each sampling time in your culture(s), calculate the mean flagellar length, both in ocular micrometer units and in micrometers. Enter the values on Data Sheet 13.1

2. On Data Sheet 13.2, record the mean values (in μm) obtained by your team for the three culture conditions and for the nondeflagellated culture.

3. On Data Sheet 13.3, record the combined mean values obtained by the entire class for the three culture conditions and for the nondeflagellated culture.

4. On Graph 13.1, plot flagellar length (ordinate) versus time (abscissa) for the three culture conditions using the combined class data. Draw each curve with a ruler and French curve template and label each plot.

5. In the Medium I culture, do the flagella attain full length by the end of the

   experiment? _____ In the Medium I culture, would you describe the rate of flagellar regeneration as acceleratory, deceleratory, or constant? Explain.

   _____

   _____

6. From the data, what is the evidence that colchicine prevents the addition of tubulin dimers onto existing microtubules?

   _____

   _____

   _____

   _____

7. How do the early measurements (0–20 min) for the Medium I and cycloheximide cultures compare?

   _____

   Explain.

   _____

   _____

   _____

8. How do the later measurements for the Medium I and cycloheximide cultures compare?

   _____

Explain.

_____

_____

_____

_____

NAME _____ SECTION _____ DATE _____

LAB PARTNERS _____

Data Sheet 13.1

Flagellar length measurements for the *Chlamydomonas* culture incubated in _____ (Medium I, colchicine, or cycloheximide) and in the nondeflagellated culture.

Sampling time (min)	Flagellar length measurements (ocular micrometer units)	Mean Mean (μm)
0		
10		
20		
30		
40		
50		
60		
75		
90		
105		

Sampling time (min)	Flagellar length measurements in the nondeflagellated culture	Mean Mean (μm)
0		
105		

Data Sheet 13.2
Team data for mean flagellar length (μm).

Sampling time (min)	Medium I ($N = 20$)	Colchicine ($N = 20$)	Cycloheximide ($N = 20$)	Nondeflagellated ($N = 20$)
0				
10				—
20				—
30				—
40				—
50				—
60				—
75				—
90				—
105				

Data Sheet 13.3
Combined class data for mean flagellar length (μm).

Sampling time (min)	Medium I (N =)	Colchicine (N =)	Cycloheximide (N =)	Nondeflagellated (N =)
0				
10				—
20				—
30				—
40				—
50				—
60				—
75				—
90				—
105				

Graph 13.1

Plot of flagellar length versus time for the cultures incubated in Medium I, colchicine, and cycloheximide.

MITOSIS AND THE CELL CYCLE

INTRODUCTION

Cell division is the process by which eukaryotic cells reproduce. The hallmark of the process is the apportionment of an identical genome to each daughter cell. Division of the nuclear contents is termed *mitosis*. Nuclear division is usually followed by cytoplasmic division, or *cytokinesis*, the apportionment of the cytosol and cytoplasmic organelles to the daughter cells as cell boundaries are formed. In this project, you will study mitosis in root tip cells of the broad bean, *Vicia faba*. It is a convenient and very popular system for studying mitosis because the chromosomes are very large, the diploid number is relatively low (2n = 12), and cell material is easily prepared. You will make squash preparations of root tip cells stained with the *Feulgen reaction* and identify all the stages of mitosis.

THE CELL CYCLE

In the life of a eukaryotic cell, mitosis is part of a larger program, the *cell cycle* (Fig. 14.1). Actively dividing cells, such as in meristematic tissues of plants or in embryonic tissues, are alternately in mitosis and in *interphase*. It is during interphase that cell growth and DNA synthesis occur, the latter during the *S* period. There are two other periods in interphase, G_1 (gap 1), which precedes DNA synthesis, and G_2 (gap 2), which follows it. A cell in G_1 is considered to have the *2C* level of DNA. During the S period, the amount of DNA doubles to the *4C* level and remains there until the end of division. After cytokinesis, each daughter cell ends up with the *2C* level of DNA. A mature eukaryotic cell that is destined not to undergo further division is usually arrested prior to DNA synthesis, in G_1. There is considerable variation both in the relative and actual duration of each stage depending on the organism, cell type, and environmental conditions. In meristematic cells of *V. faba* root tips, the cell cycle lasts about 19 hours. The specific durations of G_1, S, G_2, and mitosis have been determined from an autoradiographic analysis (Evans and Scott, 1964).

MITOTIC STAGES

Diagrams of the mitotic stages in a typical plant cell are shown in Figure 14.2. In the interphase nucleus, the chromosomal material is dispersed as a rather homogeneous mass of fibers, referred to as *chromatin*. There may be several clumps of condensed, deeply stained *heterochromatin*, and there is always at least one *nucleolus*. In your preparations, nucleoli will appear as large, lightly stained, round or oval structures

185

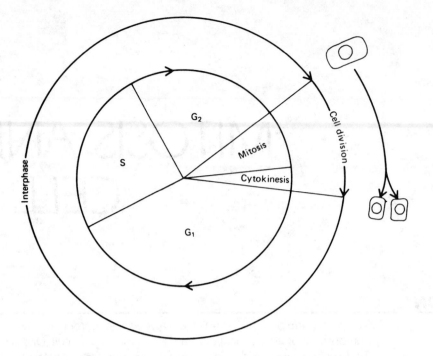

FIGURE 14.1
The cell cycle.

within the nucleus. The stages of mitosis are *prophase, metaphase, anaphase,* and *telophase.* While each mitotic stage is distinct, one stage does flow into the next so that the adjectives *early* and *late* are often appropriate.

Prophase. Prophase begins when the individual chromosomes can first be discerned. Though the chromosomes are long and thin, the double nature of each is apparent. The two subunits, termed *chromatids,* are joined at a narrow chromosomal region known as the *centromere* or *primary constriction.* Throughout prophase, the chromosomes condense and shorten. The spindle assembles in late prophase, and the chromosomes migrate toward a median position between the poles, a process termed *congression.* These movements are effected by spindle fibers extending from the poles to the chromosomes. The spindle fibers, which are composed of microtubules, terminate within the *kinetochore,* a platelike structure at the centromeric region of each chromatid. Other spindle fibers extend from pole to pole. Also occurring at late prophase is the dissolution of the nucleoli and nuclear envelope.

Metaphase. At metaphase, the fully condensed chromosomes are lined up on the *metaphase plate,* a plane that lies midway between the poles. Although all centromeres lie on the metaphase plate, the chromosome arms are still quite long and extend laterally to either side of the plate. Each metaphase chromosome has a characteristic morphology determined by its size and centromere location (Fig. 14.3). *Metacentric* chromosomes have a median centromere and two arms of equal length. In *submetacentrics,* the centromere is closer to one end. In *acrocentrics,* the centromere is very close to one end, while in *telocentrics* it is at the tip. In *V. faba* there are five pairs of acrocentric chromosomes and one pair of metacentric chromosomes; each metacentric is about twice as long as each acrocentric. Another morphological feature apparent in certain chromosomes is the *secondary constriction,* shown in the chromosome at the far right in Figure 14.3. The secondary constriction is the very thin, lightly stained chromosomal region where a nucleolus forms at telophase. The chromosomal region distal to the secondary constriction is termed a *satellite.* In *V. faba* there is a secondary constriction in one pair of chromosomes.

Anaphase. Anaphase is the climax of mitosis. The two chromatids part at the centromere and move to opposite poles, and, in addition, the poles move farther apart.

FIGURE 14.2
Mitotic stages in a typical plant cell. (From C. P. Swanson and P. L. Webster, 1977, *The Cell*, 4th ed., p. 166. Adapted by permission of Prentice-Hall, Inc., Englewood Cliffs, N.J.)

Interphase Prophase Late prophase

Metaphase Early anaphase Anaphase

Early telophase Telophase Daughter cells

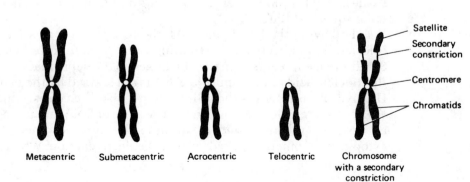

FIGURE 14.3
Morphology of metaphase chromosomes.

Metacentric Submetacentric Acrocentric Telocentric Chromosome with a secondary constriction

Satellite
Secondary constriction
Centromere
Chromatids

The mechanism of anaphase movement is not fully understood but may involve spindle microtubules sliding past each other. The shape of the anaphase chromosome depends on the location of the centromere. The centromere leads the way to the pole so that a metacentric chromosome is V-shaped, a submetacentric J-shaped, and a telocentric rod-shaped. At late anaphase, the first sign of cytokinesis occurs with the formation of the *phragmoplast*. The phragmoplast is a collection of dense material in small Golgi vesicles associated with spindle microtubules at the equatorial region.

Telophase. Telophase commences when the two daughter sets of chromosomes reach the poles. Each daughter cell receives the same number of chromosomes as had been present in the parent cell, though each chromosome now has only one subunit. There is a general reversal of events that characterize prophase. The telophase chromosomes decondense and become "fuzzy," the spindle fibers disassemble, and the nucleoli and nuclear envelope reassemble, the latter from membranes of the endoplasmic reticulum. Cytokinesis proceeds with the accumulation and fusion of vesicles at the equatorial region. The fused membranes of the vesicles produce continuous plasma membranes for the adjacent daughter cells, and the contents of the vesicles form the *cell plate*, which eventually develops into the *primary cell wall*.

MITOTIC INDEX

In any population of mitotically active cells, only some of the cells are in mitosis at any one time. The percentage of dividing cells is defined as the *mitotic index*. The approximate duration of mitosis can be obtained simply by multiplying the mitotic index by the total duration of the cell cycle. You will determine both the mitotic index and the duration of mitosis in *V. faba* root tips. For such an analysis, one should examine a cell population that is mitotically active. Since not all cells in the root apical meristem are mitotically acitve, the values you obtain for the mitotic index and the duration of mitosis will be underestimates. Having an estimate for the duration of mitosis, you will then obtain estimates for the durations of the mitotic stages. These values will also be underestimates, but the *relative* duration of each mitotic stage will be correct.

FEULGEN REACTION

To stain the chromosomes of *V. faba*, you will use the Feulgen reaction, which was introduced by R. Feulgen and H. Rossenbeck in 1924. It remains an important cytochemical test for DNA as well as a useful stain for chromosomes. The first step in the Feulgen reaction, which is depicted in Figure 14.4, is treatment of the fixed root tips with hot hydrochloric acid (HCl). The HCl hydrolysis removes the purines from DNA, leaving an exposed aldehyde group at the 1' position of the depurinated deoxyriboses. An added advantage of using the Feulgen reaction for squashes of plant tissue is that the plant fibers are *macerated* (softened) by the hot HCl, allowing the cells to be squashed flatter.

Following the HCl hydrolysis, the root tips are treated with *Schiff's reagent*, a colorless liquid prepared by treating *basic fuchsin* (a red dye) with sulfurous acid. Schiff's reagent reacts with the aldehydes on the depurinated DNA (Fig. 14.4), staining this nucleic acid reddish-purple. RNA is not stained by the Feulgen reaction since the HCl hydrolysis removes the cellular RNA. Bisulfite rinses, which are normally part of the Feulgen reaction to confirm the presence of DNA, are not necessary in this project as we are only interested in staining the chromosomes. A counterstain for the cytoplasm is sometimes used with the Feulgen reaction. When the squash preparation

FIGURE 14.4
Steps in the Feulgen reaction.

is made permanent, the cells can be stained with *fast green FCF*, which stains the cytoplasm bright green.

SQUASH PREPARATIONS

The stages of mitosis are generally studied in squash preparations. Unlike sectioned material where only a part of the cell may be present, each cell in a squash preparation contains all the chromosomes. In addition, the squashing flattens the cells so that the chromosomes are more dispersed and are often in the same focal plane.

You will want to squash only the apical meristem, for it is here that the dividing cells are found. Accordingly, you will remove the nonmeristematic tissue below and above it before making the squash. The root cap is removed by gently rubbing the very tip of the root on a paper towel. To remove the nonmeristematic tissue above the apical meristem, all but the terminal 1.5 mm of the root tip is cut away with a single-edge razor blade. The tissue is softened in a drop of 45% acetic acid on a slide. Before the preparation is squashed, the tissue is tapped with the end of a glass rod to produce a cell suspension. Then, a coverslip is lowered on the preparation. After the excess liquid is removed, the cells are squashed by thumb pressure on the coverslip. A rolling motion of the thumb is used with a paper towel between the thumb and coverslip, so that the liquid pressed to the edges of the coverslip will be absorbed. It is important not to move the coverslip laterally, as this will cause the cells to fold over on themselves. For the relatively tough cells of the root tip, maximum thumb pressure will be used, and a final round of squashing will be done with the eraser end of a pencil. If the preparation is not to be made permanent, the squash can be sealed with Vaseline and examined immediately.

PERMANENT, COUNTERSTAINED PREPARATIONS

For making the squash preparation permanent, you will use the *quick-freeze method* of A. D. Conger and L. M. Fairchild. In this method, the squashing liquid (45% acetic acid) is frozen, permitting the coverslip to be pried off with a single-edge razor blade. All or most of the tissue adheres to the slide so that, after an alcohol rinse, a drop of permanent mounting medium and a fresh coverslip can be added. In this project, you will counterstain the preparation before adding the mounting medium. This is done by simply immersing the slide (with adhering tissue) in a solution of fast green. After two ethanol rinses, a coverslip is mounted with a drop of *Euparal*, which has ethanol as the solvent.

PROCEDURES

For the Feulgen reaction, you should work in teams of four. For making the squash and permanent preparation, work individually.

FEULGEN REACTION

1. Rinse the fixed root tips in several changes of distilled water. For all rinses, cover the beaker with several layers of cheesecloth to prevent the root tips from being poured out with the rinse water.
2. Using forceps, grasp the broad end of each root tip and transfer to a beaker of 1 N HCl, 60° C. Immediately rinse the forceps in water.
3. After 10 min, dilute and cool the HCl with distilled water, pour off the liquid, and rinse the root tips briefly in distilled water.
4. With forceps, transfer the root tips to a vial of Schiff's reagent,* stoppered and kept in the dark, for 30 min. (Schiff's reagent is light-labile.) Rinse the forceps in water.
5. With forceps, transfer the root tips to a beaker and rinse in several changes of distilled water. Then rinse the forceps in water. The intensely stained distal tip contains the smaller, actively dividing cells.

SQUASH PREPARATION

1. Place a drop of 45% acetic acid in the center of a labeled, alcohol-cleaned slide.
2. Pick up a root tip by the broad end, with forceps, and gently rub the tip on a piece of paper towel to remove the root cap. Then immediately place the root tip in the drop of acetic acid.
3. With a single-edge razor blade, cut off and discard all but the terminal 1.5 mm, which contains the dividing cells. Allow the tissue to remain in the acetic acid for about 1 min.
4. Tap the root tissue with the flat end of a glass rod until there is a homogeneous suspension in the drop of acetic acid. Then, gently lower a coverslip on the preparation.
5. Before squashing, remove the excess liquid as follows. Place a paper towel on the slide, hold the towel near one end of the slide, and then move the index finger of the other hand across the slide, pressing lightly. *Be very careful not to move the*

***CAUTION:** *Be careful with Schiff's reagent. Although it is a colorless liquid, it will stain skin, clothing, and other surfaces purplish-red.*

MITOSIS AND THE CELL CYCLE

coverslip laterally, as this will cause the cells to fold over on themselves, ruining the preparation.

6. With a paper towel held in place on the slide, squash the preparation with the thumb. Use a rolling motion so that the liquid is pressed to the edges of the coverslip where the towel can absorb it. Thumb pressure should be increased in successive rounds of squashing and not applied all at once. After applying maximum pressure with the thumb, carry out a final round of squashing with the eraser end of a pencil. *Again, be careful not to move the coverslip laterally.*

7. If the slide is not going to be made permanent, seal the edges of the coverslip with Vaseline.

PERMANENT, COUNTERSTAINED PREPARATIONS

Always use forceps when placing slides in or removing slides from Coplin jars.

1. Place the slide on a flat cake of dry ice* for 5 min.

2. Pry off the coverslip with a single-edge razor blade and *immediately*, before the preparation thaws, immerse the slide in a Coplin jar of 95% ethanol for 1 min. All or almost all the tissue should adhere to the slide surface when the coverslip is removed. Keep track of which surface contains the tissue.

3. Pass the slide through a second 1-min rinse in 95% ethanol and then immerse it for 10 sec in the counterstain, a 0.5% solution of fast green in 95% ethanol.

4. Rinse the counterstained slide in two changes of absolute ethanol, 1 min each with occasional agitation, and then mount a fresh coverslip with *Euparal*. Lower the coverslip slowly to minimize air bubbles and then press out the excess liquid with a paper towel.

REFERENCES

Evans, H. J. and Scott, D. 1964. Influence of DNA synthesis on the production of chromatid aberrations by X rays and maleic hydrazide in *Vicia faba*. *Genetics 49*:17–38.

Gahan, P. B. 1984. *Plant Histochemistry and Cytochemistry*, pp. 35–37, 197–198. Academic Press, Orlando, Fla.

Nicklas, R. B. 1971. Mitosis. In *Advances in Cell Biology*, Vol. 2, Prescott, D. M., Goldstein, L., and McConkey, E., eds., pp. 225–297. Appleton-Century-Crofts, New York.

Swanson, C. P., Merz, T., and Young, W. J. 1981. *Cytogenetics*, 2nd ed., pp. 184–199. Prentice-Hall, Englewood Cliffs, N.J.

***CAUTION:** Dry ice is extremely cold and can injure the skin on contact.*

my
cells $I_m = \dfrac{17}{500} = 3.4\%$

$\dfrac{24}{403} = 5.96\%$

prepared $I_m = \dfrac{24}{403} =$

||||||| ||| ||||||

2
96
85
64   65
77
32
64
———
418
 13
———
431
  4
———
435

500 −

OBSERVATIONS AND QUESTIONS

1. Examine the preparation with the high-dry and oil-immersion objectives. Identify and draw (on pp. 197 and 199) a cell at interphase, prophase, metaphase, anaphase, and telophase. For prophase, you need draw only a portion of the nucleus, for example, a wedge-shaped section. Cells at metaphase and anaphase should have the visible secondary constrictions labeled. Record the plate magnification of each drawing.

2. Determine whether the secondary constriction is located in an acrocentric

   chromosome pair or in a metacentric chromosome pair. _____

   _____ What is the maximum number of secondary constrictions that

   would be visible in a cell at anaphase? _____
   Locate such a cell and draw it, labeling the secondary constrictions.

3. Examine several interphase nuclei, noting the prominent nucleoli. How many

   nucleoli are there per nucleus? _____ Describe the relationship
   between the number of nucleoli in a cell and the size of each nucleolus.

   _____

   What does the staining response of the nucleolus suggest about its chemical composition?

   _____

4. Score 500 cells and determine the number of cells in mitosis and interphase.

   Number of cells in mitosis = __*17*__; number of cells in interphase = __*483*__.

5. Total the number of cells in mitosis and interphase for the class. Number of cells

   in mitosis = _____; number of cells in interphase = _____. Using the

   combined class data, calculate the mitotic index. _____

6. From the value of the mitotic index and the known duration of the cell cycle in *V.*

   faba (19 hr), estimate the duration of mitosis. *38.76* min

$$\frac{17}{500} = \frac{11}{19}$$

7. Score the relative frequency of each mitotic stage in 100 dividing cells. Enter the numbers and percentages on Data Sheet 14.1. A *combined* figure should be used for anaphase and telophase, since late anaphase and early telophase are difficult to distinguish. Also enter the combined class data on Data Sheet 14.1. Using the combined class data, calculate the durations of these stages. Enter the values on Data Sheet 14.1.

8. In the Feulgen reaction, occasionally the researcher will process a control slide *not* treated with HCl. From your knowledge of the chemical basis for the Feulgen reaction, predict how such a cell preparation should stain.

_____

Explain.

_____

_____

_____

What would be the purpose of such a control?

_____

_____

13 μm

interphase

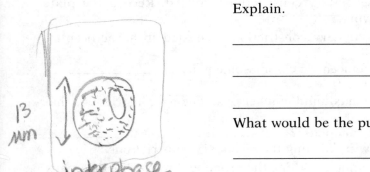

20 μm

prophase

18 μm

metaphase

22 μm

anaphase

Telophase

MITOSIS AND THE CELL CYCLE

Data Sheet 14.1

Number and percentage of cells in stages of mitosis and the estimated duration of the stages.

	Prophase	Metaphase	Anaphase + telophase	Total
Number of cells	26	8	16	~~100~~ 50
Percentage	52%	16%	32%	100%
Number of cells (class data)				
Percentage (class data)				100%
Estimated duration (min)	20.16 min	6.2	12.4	

P ||||||||| | | |||||| ||||||| ||||||

M ||||||||

A+T || ||||||| ||||||||| ||||

4,2,3 1 4 2 3 — 24

4
46
40   46
58   43
45
48
23
15
34
27
43

379

MEIOSIS

INTRODUCTION

Meiosis is a basic feature in the life cycle of all sexually reproducing eukaryotes. In a series of two divisions, the diploid number of chromosomes is reduced to the haploid number as the sperm and eggs are formed. Without this reduction there would be a progressive doubling of the chromosome number in successive generations at fertilization. Besides ensuring that the species' chromosome number remains constant, meiosis generates the bulk of the genetic diversity in nature by independent assortment of the chromosomes and crossing over. Considerable knowledge about meiotic events has accrued from studies of spermatogenesis in the grasshopper, the subject of the present project. You will make squash preparations of the grasshopper testis and study the meiotic stages.

THE TWO MEIOTIC DIVISIONS

The cells that undergo meiosis in the male derive from special cells in the testis, the *spermatogonia*. Prior to meiosis, the spermatogonia increase their numbers by mitosis. After the last mitotic division, DNA synthesis occurs, but there is a change in cell programming with the cell emerging from interphase in prophase of the *first meiotic division (M I)*. The cell during M I is termed a *primary spermatocyte*. The first meiotic division is often referred to as the *reductional division*; the homologous chromosomes pair and then segregate. The two daughter cells are termed *secondary spermatocytes*. There is no interphase between the two meiotic divisions, and there is no DNA synthesis. In some organisms there may be a slight uncoiling of the chromosomes and a pause in events termed *interkinesis*.

The *second meiotic division (M II)* is often referred to as the *equational division* because it resembles mitosis. In each secondary spermatocyte, the two chromatids of each chromosome separate and enter the nuclei of the *spermatids*. The four spermatids then differentiate into motile sperm. The transformation, which is termed *spermiogenesis*, involves growth of a flagellum and pronounced changes in the Golgi complex, mitochondria, and other organelles.

While the nuclear events of meiosis are similar in the two sexes, the cytoplasmic events are quite different. During oogenesis there is an unequal distribution of the cytoplasm at both meiotic divisions. During M I, the *primary oocyte* divides to form a *secondary oocyte*, which contains most of the cytoplasm, and a small *first polar body*. At the end of M II, the secondary oocyte gives rise to an ovum and a *second polar body*. The polar bodies degenerate so that only one functional gamete is formed.

Leptonema

Zygonema

Pachynema

Diplonema

Diakinesis

Prophase I

Metaphase I

Anaphase I

Prophase II

Metaphase II

Anaphase II

Spermatids

FIGURE 15.1
Stages of spermatogenesis
in an organism with three
pairs of autosomes and a
single X chromosome. The
paternal chromosomes are
dark and the maternal are
light.

SPERMATOGENESIS IN THE GRASSHOPPER

Figure 15.1 depicts the essential features of grasshopper spermatogenesis. For
simplicity, the figure depicts the cells of a hypothetical male organism with 7
chromosomes (3 pairs of autosomes and an X chromosome), fewer than would be
found in grasshoppers. In many grasshopper species there are 11 pairs of autosomes
and an XX-XO sex-determining mechanism. Thus, there are 24 chromosomes in the
female and 23 in the male, usually all acrocentric or telocentric. Perhaps the most
striking feature of grasshopper spermatogenesis is the behavior of the single,
heterochromatic X chromosome; during the early stages of meiosis, the X is much
more condensed and deeply stained than the autosomes. A description of the meiotic
events as they occur in grasshopper spermatogenesis is given next.

MEIOSIS I

Prophase I.   Prophase of the first meiotic division is much more complex and lasts much longer than mitotic prophase. There are five recognizable stages.

Leptonema.   The first stage of prophase I is *leptonema* (thin-thread stage). Each chromosome is very long and has a distinctive pattern of *chromomeres*, localized regions of compacted chromatin. Although there is the 4C level of DNA, only one subunit is visible in each chromosome. Frequently, the chromosomes assume a polarized arrangement, with the telomeres (chromosome tips) attached to the nuclear envelope in the vicinity of the centrioles. This *bouquet* orientation may facilitate the initial pairing of the homologues.

Zygonema.   Pairing or synapsis occurs in *zygonema* (joined-thread stage). Synapsis can begin anywhere along the chromosomes but probably occurs in a consistent manner for a given chromosome pair. Pairing proceeds in a zipperlike fashion, with the chromomeres and other structural features of the homologues aligning in precise register. All the while, the autosomes are becoming more condensed.

Pachynema.   When pairing is complete, *pachynema* (thick-thread stage) begins. The homologues are much shorter and thicker and are tightly coiled about one another. Each pair of homologues is now referred to as a *bivalent*; the unpaired X, which is still much more condensed than the autosomes, is referred to as a *univalent*. It is during pachynema that the exchange of genetic material, or crossing over, occurs. By late pachynema, the bouquet orientation has disappeared.

Diplonema.   In *diplonema* (double-thread stage), the pairing forces lapse and, as a result, each bivalent opens out. As the name of this stage implies, the double nature of each chromosome is now apparent. Another feature of diplonema is the "fuzzy" appearance of each chromatid. The two chromatids of each homologue are termed *sister chromatids*, and the chromatids of nonhomologues, *nonsister chromatids*. At one or more points along each bivalent there are cross-shaped connections between nonsister chromatids, each termed a *chiasma* (pl., *chiasmata*). Each chiasma represents a point of genetic exchange. Were it not for the presence of at least one chiasma per bivalent, the paired homologues would separate prematurely.

Diakinesis.   *Diakinesis* is the last stage of prophase I. The chromosomes become more condensed and less fuzzy, though the transition between late diplonema and early diakinesis is not always clear. As diplonema and diakinesis progress, the location of some chiasmata may change due to repulsion of the homologues and, particularly, of their centromeres. A chiasma that initially was in the middle of an arm may move or slip along the bivalent toward the tip, a phenomenon called *terminalization*. Figure 15.1 depicts several typical bivalent shapes. The final morphology of the bivalent depends on the length of the chromosome, the number and initial locations of chiasmata, and the extent of terminalization. In late diakinesis, the nucleoli and nuclear envelope disassemble, a spindle forms, and the paired homologues (bivalents) move toward the metaphase plate.

Metaphase I.   Following congression to the metaphase plate, the centromeres of each homologous pair are oriented toward opposite poles. The homologues appear to be under great tension as they stretch in the plane of the spindle. At *metaphase I*, the bivalents are lined up on the plate and exhibit very little movement. The X-univalent, on the other hand, is frequently off the metaphase plate, closer to one pole. The unorthodox location of the X results from its repeated pole-to-pole movements while the autosomes remain lined up on the plate. Also, at metaphase I, the X is usually more lightly stained than the autosomes. The arrangement of the homologous pairs is random so that independent assortment occurs; either homologue (maternal or paternal) is as likely to face the same pole and, hence, enter the same daughter nucleus as either homologue of any other bivalent. So, in Figure 15.1, for example, the dark (paternal) long chromosome enters the same daughter nucleus as the light (maternal) short chromosome, but it would be just as likely for *both* long and short maternals (or paternals) to go to the same pole. The independent assortment and crossing over together generate the genetic variability that the process of meiosis brings about.

Anaphase I and Telophase I.   At *anaphase I* there is synchronous separation of the homologues. The number of arms in each half-bivalent depends on the location of the centromere, which leads the way to the pole. Telocentric and acrocentric chromosomes have two arms, while metacentrics and submetacentrics have four arms. Following *telophase I*, each of the two nuclei has the haploid number of chromosomes, and each has the 2C level of DNA.

MEIOSIS II

The second meiotic division very much resembles mitosis with three noteworthy differences at M II: (1) the haploid number of chromosomes is present, (2) the two chromatids of each chromosome are not genetically identical, and (3) the two chromatids are not tightly coiled about each other but are widely separated both at *prophase II* and *metaphase II*. At metaphase II, the chromosomes are lined up midway between the poles (Fig. 15.1), but in squash preparations they are often observed to be in a circle, an orientation that results when the cell is squashed with the spindle pole at the center. At *anaphase II*, the chromatids of each chromosome separate. As they move to the poles, telocentrics appear as rods, while metacentrics have two arms. At the end of *telophase II*, each nucleus is haploid, each has the 1C level of DNA, and each is genetically unique.

SUPERNUMERARY OR B-CHROMOSOMES

In many grasshopper populations, males are found with more than 23 chromosomes. Such individuals contain *supernumerary* or *B-chromosomes*. B-chromosomes, which are usually small and heterochromatic, are additional to the regular chromosomes (A-chromosomes) and are not homologous with them. Curiously, the B-chromosomes have only a slight effect on the phenotype, even when present in several doses.

SQUASH PREPARATIONS

Meiosis is generally studied in a squash preparation (see Project 14, p. 189). You will be making squash preparations of testis material that has previously been fixed in acetic-ethanol and stained with alcoholic HCl-carmine. After transferring the testis to 70% ethanol, you will tease apart the *follicles*, the tubular sacs that comprise the testis. You will then squash the follicles in 45% acetic acid. Since the cells in each follicle exhibit a degree of synchrony, you will probably have to examine more than one slide to locate all the stages.

PROCEDURES

TESTIS SQUASH PREPARATION

1. With forceps, transfer a stained testis, or part of one, to a Syracuse dish containing 70% ethanol.

2. Using a dissecting microscope and dissecting needles, tease apart several follicles from the testis. Place a drop of 45% acetic acid on an alcohol-cleaned slide and place three follicles in the drop. *The tissue must not be allowed to dry out.* Replace the cover on the Syracuse dish.

3. With the dissecting needles, cut each follicle in half. Remove and discard the tapered half, since the dividing cells will be in the portion with the blunt end. Spread out the remaining pieces of tissue and add a coverslip.

4. Before squashing, remove excess liquid as follows. Place a paper towel over the slide, hold the towel near one end of the slide, and then move the index finger of the other hand across the slide, pressing lightly. *Be very careful not to move the coverslip laterally*, as this will cause the cells to fold over on themselves, ruining the preparation.

5. With a paper towel held in place on the slide, squash the preparation with the thumb. Use a rolling motion so that the liquid is pressed to the edges of the coverslip where the towel can absorb it. Thumb pressure should be increased in successive rounds of squashing and not applied all at once. After applying maximum pressure with the thumb, carry out a final round of squashing with the eraser end of a pencil. *Again, be careful not to move the coverslip laterally.*

6. Check the slide for the presence of dividing cells. If meiotic stages are present, the edges of the coverslip can be sealed with Vaseline and the slide examined. Alternatively, the preparation can be made permanent using the procedure given in Project 14, page 191, except that it should *not* be counterstained. After the coverslip is removed from the frozen preparation, rinse the slide in 95% ethanol and absolute ethanol, 1 min each, and then mount a coverslip with Euparal.

REFERENCES

Macgregor, H. and Varley, J. 1983. *Working with Animal Chromosomes*, pp. 39–69. John Wiley, New York.

Stephens, R. T. and Bregman, A. A. 1972. The B-chromosome system of the grasshopper *Melanoplus femur-rubrum*. *Chromosoma* 38:297–311.

Swanson, C. P., Merz, T., and Young, W. J. 1981. *Cytogenetics*, 2nd ed., pp. 200–233. Prentice-Hall, Englewood Cliffs, N.J.

White, M. J. D. 1973. *Animal Cytology and Evolution*, 3rd ed., pp. 148–197; 312–333. Cambridge Univ. Press, London.

OBSERVATIONS AND QUESTIONS

1. Identify and then draw (on pp. 209–213) the following stages: pachynema, diplonema, diakinesis, metaphase I, anaphase I, metaphase II, anaphase II, and several intermediates in spermiogenesis. Label the X-univalent in the prophase drawings and at metaphase I. Record the plate magnification of each drawing.

2. How can you recognize the X chromosome at anaphase I?

_____

_____

3. From your drawing of diplonema, determine the number of chiasmata in each large bivalent, each medium bivalent, and each small bivalent. Record the numbers, along with the mean for each type of bivalent.

	Number of chiasmata	Mean
Large bivalent		
Medium bivalent		
Small bivalent		

Describe the relationship between the number of chiasmata and chromosome length.

_____

_____

4. How many *different* bivalent configurations do you observe at metaphase I?

_____ In the space provided, draw two different configurations you have observed and next to each make a line drawing showing the four chromatids, the centromeres, and the location of the chiasma(ta).

5. Score the number of chromosomes in 10 cells at metaphase II and enter the numbers in the spaces provided.

	Cell									
	1	2	3	4	5	6	7	8	9	10
Chromosome number										

How do you explain the variation?

_____

_____

6. Determine whether there are any supernumerary chromosomes present.

TRANSCRIPTION OF RIBOSOMAL RNA

INTRODUCTION

In eukaryotes, most ribosomal RNA (rRNA) is transcribed at the *nucleolar organizer region*, or *NOR*, on one or more chromosome pairs. The NOR is located in the secondary constriction and, as its name implies, is the chromosomal site where the nucleolus forms at telophase. Within the nucleolus, the ribosomal subunits are assembled from the transcribed rRNA and ribosomal proteins. In the human chromosome complement, there is an NOR on five pairs of chromosomes, although not all the NORs always function in nucleolar synthesis. In this project, you will become familiar with the human chromosome complement by examining slides that are conventionally stained with Giemsa blood stain. You will also use a silver staining reaction on human chromosomes from cultured cells to identify the transcriptionally active NORs.

rRNA TRANSCRIPTION

The genes that code for ribosomal RNA comprise only 1–2% of the eukaryotic genome, yet, typically, 80–90% of the total cellular RNA is rRNA. The large quantity of rRNA results from the presence of many copies of the rRNA gene as well as from their continual transcription during interphase. In a typical eukaryotic cell, each NOR contains several hundred copies of the rRNA gene arranged in tandem. The DNA in each rRNA gene is termed *ribosomal DNA (rDNA)*. Transcriptionally active rDNA has been visualized with the electron microscope. The first such observations were reported in 1969 by O. L. Miller and B. R. Beatty, who studied the chromatin in dispersed nucleoli isolated from amphibian oocytes. Figure 16.1 is a diagrammatic representation of the structure observed. Each "fern leaf" configuration consists of a core of rDNA, also termed a *transcriptional unit*, and a matrix of growing RNA

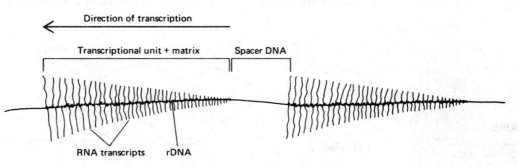

FIGURE 16.1
Diagrammatic representation of transcriptionally active rRNA genes as they appear in electron micrographs.

transcripts. The start of each transcriptional unit is located at the tip of the fern leaf, where the RNA transcripts are shortest, and the end of transcription is at the broad end, where they are longest. Between adjacent transcriptional units there is nontranscribed *spacer DNA*. The completed RNA transcript is by no means the final form of the rRNA as found in the ribosome. There is further processing of the RNA associated with the self-assembly process that gives rise to the ribosomal subunits, as described next.

THE RIBOSOME AND ITS ASSEMBLY

The eukaryotic ribosome is a small, oblong particle, about 22 nm × 32 nm. The intact ribosome, its subunits, and the various classes of rRNA are designated by their sedimentation coefficients in Svedberg units (S). A discussion of sedimentation coefficients can be found in Project 8, page 000. The intact eukaryotic ribosome (Fig. 16.2) is an 80S particle, separable into two subunits, a small or 40S subunit and a large or 60S subunit. The 40S subunit contains a molecule of 18S rRNA and about 30 proteins, while the 60S subunit contains three molecules of rRNA (28S, 5.8S, 5S) and about 40 proteins.

The RNA molecule produced at the NOR is a large 45S transcript. In the nucleolus, the 45S transcript is sequentially degraded into the 18S, 28S, and 5.8S rRNAs. The 5S rRNA is synthesized elsewhere in the genome, and the numerous ribosomal proteins are synthesized in the cytoplasm. After their synthesis, the 5S rRNA and ribosomal proteins migrate to the nucleolus. In the nucleolus, the various rRNAs and ribosomal proteins undergo a complex assembly process giving rise to the 40S and 60S ribosomal subunits. When their assembly is complete, the 40S and 60S subunits leave the nucleolus and migrate to the cytoplasm. There, in association with messenger RNA and transfer RNA, the subunits unite to form the 80S particle, which functions in protein synthesis.

AMMONIACAL SILVER STAINING

The staining reaction that reveals chromosomal sites of prior rRNA transcription is the *ammoniacal silver* method, abbreviated *Ag-AS*. The procedure was developed by Goodpasture and Bloom in 1975 and later improved by these same workers in 1976. What is observed is a deeply stained NOR contrasted against the lightly stained chromatin in the remainder of the chromosome.

Evidence suggesting that the ammoniacal silver procedure stains only transcriptionally active NORs derives from studies by D. A. Miller, O. J. Miller, and their associates. Their experiments involved analyses of rRNA and chromosomal staining in *mouse–human hybrid cell lines*. These somatic cell hybrids were produced by culturing mouse and human cells together in the presence of inactivated Sendai virus, which induces cell fusion. Some of the hybrid cell lines contained only mouse 28S rRNA, while other hybrid cell lines contained only human 28S rRNA. In cells where only the mouse rRNA had been synthesized, only the NORs in the mouse chromosomes stained with the Ag-AS method, even though NOR-containing human chromosomes were present. Conversely, in cells where only the human rRNA had been

FIGURE 16.2
Components of the eukaryotic ribosome.

synthesized, only the human NORs stained. Although the cytochemical basis for the silver staining is not completely understood, it appears that the stained material at the NOR is protein. Goodpasture and Bloom have found that prior treatment with trypsin or pronase eliminates the staining of the NORs.

The ammoniacal silver method is relatively simple. The procedure is carried out on a slide preparation of dividing cells. First, the slide preparation is incubated with a hot solution of silver nitrate. This is followed by a short exposure to a combined mixture of ammoniacal silver plus a formalin developer. The progression of the staining is monitored microscopically, for the reaction must be halted at just the right moment. The NOR rapidly blackens, while the remainder of the chromosome gradually becomes yellow or brown. If the staining proceeds too far, the entire chromosome becomes dark brown or black. Preparations that are properly stained have black NORs contrasted against golden yellow or light brown chromosomes. When the correct degree of development is observed, the staining reaction is stopped by immediately rinsing the slide in water. No other staining is required. The last step is the mounting of a coverslip with Permount.

SHORT-TERM LEUKOCYTE CULTURES

The source of the chromosomes for this investigation is the short-term human leukocyte culture. Such cultures are initiated with a few drops of whole blood in the growth medium and yield a large number of dividing cells in only several days. Like all mammalian cells in culture, leukocytes require a rich growth medium. Human leukocytes are grown in a medium containing fetal calf serum, amino acids, vitamins, buffered salts, and antibiotics, the latter to inhibit bacterial growth. The leukocytes of normal individuals will not undergo division unless the medium also contains a *mitogen*, a substance that stimulates the cells to divide. The most commonly used mitogen in leukocyte cultures is *phytohemagglutinin*, a mucoprotein derived from the red kidney bean. After 24 hours in culture, the small lymphocytes commence DNA synthesis and subsequently undergo mitosis. Peak mitotic activity occurs at some time between 60 and 72 hr.

Metaphase Arrest.   For chromosome studies, we like the chromosomes to be relatively condensed, as they are at metaphase. To accumulate a large number of metaphases, the plant alkaloid *colchicine* is added for the final 2–3 hr of culture. Colchicine interferes with spindle formation so that dividing cells do not proceed through mitosis but are arrested at a stage termed *c-metaphase* (*c* is for colchicine).

CELL HARVEST

The basic steps in the harvest are brief treatment with a hypotonic solution, fixation of the cells, and air drying of the fixed cell suspension on slides. The cells are collected from the culture by centrifugation, washed in a balanced salt solution, and then treated with a hypotonic solution of potassium chloride (KCl). The hypotonic treatment is a critical step, for it swells the cells and disperses the chromosomes. After removal of the hypotonic KCl, the cells are fixed in 1:3 acetic-methanol, a rapidly penetrating and volatile fixative. Several drops of the suspension of fixed cells are placed on a microscope slide, which is quickly dried. The rapid drying results in well-spread metaphases in which all the chromosomes are in one focal plane. You will receive slide preparations that are ready for silver staining.

THE HUMAN KARYOTYPE

Before you can study the nucleolar organizer regions with and without silver staining, you will need to become familiar with the morphology of human chromosomes. The normal human chromosome complement consists of 22 pairs of autosomes and a pair of sex chromosomes, XX in females and XY in males. The *karyotypes* of a normal human female and male are shown in Figure 16.3. A karyotype is an ordered arrangement of the chromosomes, with their placement determined using standard rules agreed upon by cytogeneticists. The chromosome pairs are serially numbered by decreasing length and are placed into seven groups (A–G) by centromere location or *arm ratio*, (length of long arm)/(length of short arm). On the basis of size and arm ratio, every chromosome can unequivocally be placed into one of the seven groups. Certain chromosomes can be identified individually with conventional staining. In Figure 16.3, these chromosomes have a number directly beneath them; the other chromosomes are designated by group only. For positive identification of every human chromosome, *chromosome banding* procedures are used. These staining methods produce a distinctive pattern of transverse dark and light bands along the length of each chromosome.

The D and G Chromosomes.   In this project, you will be concerned with the D (13–15) and G (21 and 22) chromosomes. These are the human chromosomes that contain the NORs, located in the secondary constrictions. The G chromosomes are the smallest (about 2 μm) in the human chromosome complement. The D chromosomes are appreciably longer than the G chromosomes but otherwise have a similar morphology, as can be seen in Figure 16.3. All the D and G chromosomes are acrocentric with a secondary constriction and satellite in the short arm (see discussion of chromosome morphology in Project 14, p. 186). Although shown in the stylized karyotypes (Fig. 16.3), the secondary constriction and satellite are usually not apparent in every D and G chromosome in the same cell.

Satellite Association.   An analysis you will perform using the Giemsa-stained slides relates to the arrangement of the D and G chromosomes within the metaphase spread. More frequently than would be expected by chance, the short arms of two or more D and/or G chromosomes lie close to each other and are directed toward each other. This arrangement, which is termed a *satellite association*, appears to be related to the common role of the D and G chromosomes in nucleolar synthesis.

PROCEDURES

SILVER STAINING (Ag-AS Method)

You should stain your own slide preparation but share the various solutions with one other person. Work with only one slide at a time. Since the silver solutions can stain skin and clothing, it is imperative that a lab coat and disposable gloves be worn. The lab bench and the floor beneath it should be covered with Benchkote and a drop cloth.

1. Place a labeled slide in a Petri dish. With a Pasteur pipet, flood the slide surface with a freshly filtered aqueous solution of 50% silver nitrate. Add a large coverslip and place the covered Petri dish in a 50°–60° C oven for 15 min.
2. Rinse the slide thoroughly under running distilled-deionized water and allow it to air dry.
3. Examine the slide under low power with the condenser iris closed and locate a region where there are several good metaphase spreads. Confirm your observations under high-dry and then carefully remove the slide from the stage without disturbing the stage coordinates; you will be monitoring the progress of the staining in these metaphase spreads.
4. Remove all colored filters from the microscope and place two Coplin jars with distilled-deionized water on your lab bench. These water rinses are required to

FIGURE 16.3
Stylized karyotypes of a normal human female (a) and male (b). (From V. A. McKusick, 1969, *Human Genetics*, 2nd ed., pp. 9, 10. Adapted by permission of Prentice-Hall, Inc., Englewood Cliffs, N.J.)

stop the staining reaction, which proceeds rapidly after you add the staining solutions, as described in step 5. When optimal staining is observed (usually 15–45 sec), the NORs will be black and the chromosome arms golden yellow or light brown. If the preparation is stained too long, the entire chromosome will blacken, rendering the NOR indistinguishable from the rest of the chromosome. To prevent overstaining, be prepared to examine the slide immediately after adding the solutions in step 5.

5. With a fresh Pasteur pipet, place four drops of ice-cold, freshly filtered ammoniacal silver solution on the slide surface followed by four drops of ice-cold formalin developer, dispensed with another fresh Pasteur pipet. Immediately add a large coverslip and place the slide on the microscope stage to follow the progress of the staining.

6. As soon as optimal staining is observed, i.e., black NORs and golden yellow or light brown chromosome arms, *immediately* remove the coverslip and immerse the slide in two changes of distilled-deionized water.

7. Rinse the slide more thoroughly in several changes of distilled-deionized water and allow it to air dry in a near vertical position.

8. Using forceps, dip the air-dried slide in xylene (in a fume hood) and mount a coverslip with two drops of Permount. Lower the coverslip slowly to minimize air bubbles and then gently press out the excess liquid with a paper towel.

REFERENCES

Bloom, S. E. and Goodpasture, C. 1976. An improved technique for selective silver staining of nucleolar organizer regions in human chromosomes. *Hum. Genet.* 34:199–206.

Goodpasture, C. and Bloom, S. E. 1975. Visualization of nucleolar organizer regions in mammalian chromosomes using silver staining. *Chromosoma 53*: 37–50.

Miller, D. A., Dev, V. G., Tantravahi, R., and Miller, O. J. 1976. Suppression of human nucleolus organizer activity in mouse–human somatic hybrid cells. *Exp. Cell Res.* 101:235–243.

Miller, Jr., O. L. 1984. Some ultrastructural aspects of genetic activity in eukaryotes. *J. Cell Sci. Suppl. 1*:81–93.

Therman, E. 1985. *Human Chromosomes*, 2nd ed. Springer-Verlag, New York.

OBSERVATIONS AND QUESTIONS

1. On the Giemsa-stained slide, locate a well-spread cell at c-metaphase. Using the oil-immersion objective, identify at least one D chromosome and at least one G chromosome that have visible secondary constrictions and satellites. Then, on page 223, draw these chromosomes, labeling the centromeres, secondary constrictions, and satellites.

2. Locate a cell in which a satellite association is evident. The association can involve any combination or two or more D and/or G chromosomes. Draw the satellite association, labeling each chromosome (D or G) and any satellites that are visible.

3. On the Ag-AS slide, locate a well-spread cell at c-metaphase in which the chromosomes are properly stained; the staining response may differ in different regions of the slide. What is the total number of silver-stained NORs (Ag-NORs)

   in the cell? _____ How many D chromosomes have an Ag-NOR? _____ How

   many G chromosomes have an Ag-NOR? _____ How do you explain the variation in the size of the individual Ag-NORs?

   _____

   _____

4. Score the total number of Ag-NORs in each of five cells. Enter these values on the lines provided.

   _____   _____   _____   _____   _____

   What is the range for the total number of Ag-NORs? _____ What is the mode

   (most frequent number)? _____

5. Is the observed number of Ag-NORs ever as great as the total number of NORs

   present in the human chromosome complement (10)? _____ How do you explain the difference?

   _____

   _____

6. Locate a cell in which there is a satellite association of chromosomes with Ag-NORs. Draw the chromosomes in the association as well as one or two other chromosomes with Ag-NORs. Would you be surprised to find a satellite

   association in which the NORs were *not* stained? _____ Explain.

   _____

   _____

7. Locate a cell at prophase, where the chromosomes are longer and less condensed. Examine the Ag-NORs and determine whether the staining is at the very tip or just beneath the tip of the chromosome. Draw a prophase chromosome with an Ag-NOR. Do your observations indicate that rRNA transcription is occurring at the satellite or at the secondary constriction? Explain.

DNA REPLICATION

INTRODUCTION

The *basic* mechanism of DNA replication was first suggested by J. D. Watson and F. H. C. Crick as part of their work on the structure of DNA. In the replication process, the two strands of the DNA double helix (or duplex) separate, and each original strand serves as a template for the synthesis of a complementary strand. The term *semiconservative* is used to describe this mode of replication, since one strand, i.e., *half* of the parental duplex, remains intact in each daughter duplex. The first evidence indicating that the DNA in eukaryotic chromosomes replicates semiconservatively was provided by J. H. Taylor and coworkers in 1957. They allowed dividing root tip cells of *Vicia faba* to incorporate ^{3}H-thymidine and then followed the labeling pattern at successive metaphases using the technique of autoradiography. DNA replication in eukaryotic chromosomes can also be followed cytochemically using a special dye, *Hoechst 33258*. You will use the latter methodology in this project to follow the course of DNA replication in cultured human leukocytes.

BrdU–HOECHST METHOD

Hoechst 33258 is a *fluorescent* chromosome stain; the dye molecules bind to the DNA, and the chromosomes fluoresce brightly. In 1973, S. A. Latt determined that the fluorescence is reduced if the base analogue *5-bromodeoxyuridine* (*BrdU*) is incorporated in the DNA. BrdU resembles thymine except that at the 5 position of the pyrimidine ring there is a bromine atom instead of a methyl group. Latt grew human lymphocytes in medium containing BrdU, which is incorporated in the chromosomal DNA in lieu of thymine. He then stained the cell preparation with Hoechst 33258. In cells that had undergone one replication in the BrdU medium, both sister chromatids fluoresced brightly, though not as brightly as chromosomes without *any* BrdU. After two replications in the BrdU medium, one sister chromatid fluoresced brightly and the other fluoresced weakly. Other experiments revealed that the weakly fluorescing chromatid contained about twice as much BrdU as its brightly fluorescing sister chromatid. Thus, the brightly fluorescing chromatid has BrdU incorporated in only one strand of its DNA, while the weakly fluorescing sister chromatid has BrdU incorporated in both strands. As shown in Figure 17.1a, these observations are precisely what would be predicted if an unreplicated chromosome contains one DNA duplex, and replication proceeds semiconservatively. An additional observation at the second metaphase is the presence of "harlequin" chromosomes (Fig. 17.1b). The alternating pattern of bright and dull regions along each chromatid results from *sister chromatid exchanges* (*SCEs*), i.e., the breakage and reunion of homologous segments of sister chromatids.

FIGURE 17.1

(a) Incorporation of 5-bromodeoxyuridine (BrdU) in DNA and the subsequent appearance of the chromosomes after staining with the fluorescent dye Hoechst 33258. Chromatids in which one DNA strand is substituted with BrdU fluoresce brightly (deeply shaded), whereas chromatids in which both DNA strands are substituted with BrdU fluoresce weakly (lightly shaded). (b) Formation of a chromosome with two sister chromatid exchanges (at arrows) and its appearance at the second metaphase.

HOECHST–GIEMSA STAINING

The BrdU–Hoechst technique, as done by Latt, has a technical problem in that the fluorescence fades rather quickly. This difficulty was obviated by Wolff and Perry (1974), who found that if the Hoechst-stained slide is exposed to light for 24 hr, the BrdU-substituted chromatids at the second metaphase can be differentially stained with ordinary Giemsa blood stain. With this procedure, known as the *fluorescence-plus-Giemsa (FPG)* method, the chromatid with one BrdU-substituted strand of DNA stains deeply, while the chromatid with two BrdU-substituted strands stains lightly. Apparently, the exposure to light interferes with the binding of the Giemsa dye to the doubly substituted, Hoechst-stained chromatid. Not only is the Giemsa-stained preparation resistant to fading, but it is also excellent for detecting sister chromatid exchanges. There is currently great interest in SCEs, for they are an indication of genetic damage. Accordingly, Hoechst–Giemsa staining is used for screening potential mutagens and carcinogens (Wolff, 1982).

The protocol selected for the present study was developed by Goto *et al.* (1975). Human leukocytes are cultured for two to three cell generations in medium containing BrdU. The principles underlying the culturing of human blood cells are discussed in Project 16, page 217. You will be provided with unstained slide preparations from such harvested cultures.

The first step in the FPG staining procedure is brief treatment of the unstained slide preparations with a relatively high concentration of Hoechst 33258. By using a high concentration of the dye, Goto and coworkers found that the time required for light exposure is dramatically reduced. After a 1-hr exposure to an intense fluorescent light source, the slide preparations are stained with Giemsa blood stain. The last step is the mounting of a coverslip with Permount.

PROCEDURES

HOECHST–GIEMSA STAINING

Always use forceps to place slides in or to remove slides from Coplin jars.

1. Dip a labeled slide briefly in distilled water and then place in an aqueous solution of Hoechst 33258, 200 µg/ml, for 10 min.

2. With forceps, transfer the slides to an empty Coplin jar and then rinse the slides under running tap water for 2 min.

3. Replace with distilled water, pour it off, and then prop the slides in a near vertical position until dry or nearly so.

4. With a Pasteur pipet, flood the surface of the slides with McIlvaine's buffer. Place a large coverslip on each slide.

5. Place the slides approximately 3 cm from two fluorescent bulbs. Expose the slides to the light for 1 hr, checking occasionally that the coverslips are in place.

6. Remove the coverslips and then rinse each slide under running distilled water for several seconds.

7. Stain the slides for 6 min in a freshly prepared 4% Giemsa solution. Convenient volumes for a Coplin jar are 1.5 ml Giemsa stain,* 6.0 ml phosphate buffer,* 30 ml distilled water.

8. Rinse each slide under running tap water for 5 sec and allow to air dry in a near vertical position. When each slide is completely dry, dip briefly in xylene (in a fume hood) and mount a coverslip with two drops of Permount. Lower the coverslip slowly to minimize air bubbles and then gently press out the excess liquid with a paper towel.

REFERENCES

Goto, K., Akematsu, T., Shimazu, H., and Sugiyama, T. 1975. Simple differential Giemsa staining of sister chromatids after treatment with photosensitive dyes and exposure to light and the mechanism of staining. *Chromosoma* 53:223-230.

Latt, S. A. 1973. Microfluorometric detection of deoxyribonucleic acid replication in human metaphase chromosomes. *Proc. Nat. Acad. Sci.* 70:3395–3399.

Wolff, S., ed. 1982. *Sister Chromatid Exchange.* John Wiley, New York.

Wolff, S. and Perry, P. 1974. Differential Giemsa staining of sister chromatids and the study of sister chromatid exchanges without autoradiography. *Chromosoma* 48:341–353.

*CAUTION: *Do not pipet by mouth.*

OBSERVATIONS AND QUESTIONS

1. Scan the slide for a cell at the first metaphase. How do you know that the cell has undergone only one division in the BrdU medium?

   _____

   _____

   _____

2. Scan the slide for a cell at the second metaphase. On page 231, draw several chromosomes in this spread, shading the chromatids to show their staining intensity.

3. From the cell in question 2, draw the chromosome that shows the greatest number of SCEs. How many reciprocal breaks must have occurred to produce

   the observed configuration? _____

4. Scan the slide for a cell in which some chromosomes have differentially stained chromatids and some have both chromatids lightly stained. How many

   replications in BrdU did this cell undergo? _____ Explain your answer using diagrams similar to those in Figure 17.1a.

5. Using diagrams, explain the origin of a chromosome with the following appearance:

PROJECT
18

PHOTOMICROGRAPHY

INTRODUCTION

Photomicrography is a valuable adjunct to microscopy, for it provides the cell biologist with an accurate, permanent, visual record of microscopic observations. Prior to the invention of photography, the only means of recording such images was by sketching them. In 1850, even though photography was a primitive science, acceptable photomicrographs were being made on wet plates. With the advent of dry emulsions, specially designed photomicrographic cameras, and improved microscope optics, taking quality photomicrographs became a routine matter. The visual record takes two forms, the *print* for published works and the *transparency* for projection to an audience. Prints and transparencies can be made with "instant" films or with traditional wet-process methods, which involve chemical processing in a darkroom. Included in this project are protocols for making black-and-white prints, either with the wet-process method or with instant film. Also included is a protocol for making black-and-white transparencies with instant 35mm film. You will be able to photograph a variety of preparations you have made in the cell biology laboratory.

PRINCIPLES OF PHOTOMICROGRAPHY

It is possible to record microscopic observations on film because, in addition to the virtual image perceived by the eye, there is a real image projected through the ocular. When a camera is placed above the ocular, generally the photographic ocular of a *trinocular microscope*, the image can be captured on film. Before any photomicrography is done, the microscope must be adjusted for Köhler illumination (Project 1, p. 7), with the condenser iris set for optimum contrast. Regardless of how sophisticated the camera, the quality of the photomicrograph will be less than ideal if the microscope is improperly adjusted.

The basic steps for taking photomicrographs are the same as in pictorial photography. First, the picture is composed; the cells or structures to be photographed are positioned in the field so that they are included in the film frame. Next, the image is brought into sharp focus at the film plane. Finally, the correct exposure is determined. These steps are described for two popular film formats, 35mm film and Polaroid pack film.

Composition and Magnification. The picture is composed with the aid of a *focusing eyepiece* (Fig. 18.1). This special ocular contains a reticle with one or more frames, indicating the region of the microscope field that will be included on the film frame with different size films. The 35mm frame is 24mm × 36 mm and the Polaroid print frame is 8.3 cm × 10.8 cm (3¼ in. × 4¼ in.). When the field you would like to photograph is larger than the reticle frame, the total magnification should be reduced. By turning the focusing eyepiece, you can select the regions of the field that will be

FIGURE 18.1
Focusing eyepiece and its focusing/framing reticle.
(Courtesy of Reichert Scientific Instruments.)

Labels in figure: Collar, Field of view, Eyepiece, IO X WF

included in the frame. When good composition is achieved, the camera is turned so that it is positioned at the same angle as the reticle frame.

The magnification at the film plane is determined by the particular objective and photographic ocular being used. If prints are to be made, it is desirable to photograph the calibrations of a stage micrometer slide. This negative will later be printed.

Focusing.   The focusing eyepiece enables you to focus the microscope image at the film plane of the camera. First, the collar on the focusing eyepiece is turned until the cross hairs on the reticle (Fig. 18.1) are in sharp focus. Then, the image is brought into focus with the fine adjustment knob. If you have difficulty focusing on the cross hairs, it may be helpful to gaze at a distant object before focusing.

Determining the Correct Exposure.   In all photography, the *exposure* (*E*) is determined by *light intensity* (*I*) and *exposure time* (*T*), the relationship being $E = IT$. Thus, the exposure remains the same if, for example, the light intensity is halved and the exposure time simultaneously doubled. In photomicrography, the intensity of the light that reaches the film is affected by many factors. These include the transformer setting, which determines the voltage reaching the illuminator bulb, any filters in the light path (see next section), the magnifications of the objective and ocular, and, of course, the nature of the specimen. Generally, one controls light intensity by changing the transformer setting or the filter in the light path.

The exposure time is controlled by the shutter setting. As in pictorial photography, an *exposure meter* is a great aid in determining the correct exposure time. Before obtaining the meter reading, you will have to adjust the meter for the *ASA speed* of the film you are using. The ASA speed is an arithmetic measure of the sensitivity of the film to light; the higher the ASA speed, the "faster" or more light sensitive is the film. For example, a film with an ASA speed of 100 requires one-half the exposure of a film with an ASA speed of 50. Some exposure meters must also be adjusted for the film format (35mm or 3¼ in. × 4¼ in.) and the magnification of the photographic ocular. The meter reading, hereafter termed the *indicated exposure time*, is only an estimate of the best exposure time. If the indicated exposure time is greater than the slowest available shutter speed on your photomicrographic assembly, obviously you will have to turn up the transformer setting.

To obtain the best exposure time, it is necessary, at first, to run a series of exposures with successively doubled exposure times. The exposure series should

234

bracket the indicated exposure time. Assume, for example, that the indicated exposure time is 1/15 sec. Then, a series of three 35mm frames can be exposed at 1/30, 1/15, and 1/8 sec, and you can be fairly certain of having one properly exposed frame. For Polaroid instant prints, where each frame is more costly, an exposure series can be made on *one* print if your camera back has a *dark slide* close to the film plane. The multiple exposures on the single print can then be examined to decide on the best exposure time for the final print.

After you have made an exposure series with 35mm film, you will have to determine which frame has the best exposure. Selecting which of several transparencies has the best exposure is an easy matter since it is similar to selecting a good print. Selecting the best negative in an exposure series is more difficult for the beginner because the negative has the *reverse* light–dark features of the subject. You should check to see which frame has the greatest detail in *light* areas on the negative. If the negative is underexposed, there will be very little detail in such areas. If the negative is greatly overexposed, you will not be able to see newsprint through the darkest areas of the negative. Since negative films have considerable exposure latitude, it is likely that more than one frame will be printable. However, in choosing between a slightly overexposed and a slightly underexposed negative, it is better to print the former.

Filters. To enhance contrast in black-and-white photomicrography, we use filters with colors that are complementary to those of the stained specimen. For the red and purple hues in stained cells and tissues, a green filter, such as a Kodak Wratten Filter No. 58, or a yellow-green filter, such as a No. 11, should be used. The green filtration also minimizes the effects of lens aberrations since objectives are optically corrected in this spectral region.

If the light intensity is very high, as would be the case at low magnification, the correct exposure time might be shorter than the fastest available shutter speed. Light intensity can be reduced by lowering the transformer setting or by inserting a *neutral density filter*. Commonly used neutral density filters have transmittances of 50%, 25%, and 12.5%.

Checklist. With so many factors affecting the quality of each photomicrograph, it is essential to keep a record of important variables. A checklist for your personal record is provided on Data Sheet 18.1. If you have kept a careful record, you will know the best exposure time for each combination of film, stain, and filter. Such data will be helpful for taking photomicrographs in the future.

BLACK-AND-WHITE, WET-PROCESS METHODS

35mm Film. An excellent 35mm film for routine black-and-white photomicrography is Kodak *Technical Pan Film 2415*. It has extremely high resolving power and also has extremely fine grain, permitting the negative to be printed with considerable enlargement. With the processing protocol given in this project, Technical Pan has an ASA speed of 50 and yields negatives of moderate contrast. Like most films, Technical Pan is *panchromatic*, i.e., sensitive to all wavelengths in the visible spectrum, and must be handled in complete darkness.

Film Processing. All films consist of two basic layers. There is a light-sensitive *emulsion*, which is a layer of gelatin containing crystals of silver halides, and a *base*, which is a polyester or cellulose acetate support surface. The silver halides are very light-sensitive, and even a small amount of light reduces some of the silver ions to metallic silver in small regions of the crystals. The image formed is not visible to the eye and is termed a *latent image*. To produce a *visible image*, the film must be *developed*. The developer chemically reduces more of the silver ions in the exposed crystal, producing a *grain* of metallic silver. Development is arrested with a *stop bath*, a weak solution of acetic acid that neutralizes the alkaline developer. Following development, the film is *fixed,* or made permanent, with *fixer,* which contains sodium hyposulfite and a chemical that hardens the gelatin coat. The sodium hyposulfite makes the image permanent by converting the unexposed silver halide crystals to soluble compounds, which can then be rinsed away. The finished negative has the reverse light–dark features of the subject, for there are now dark areas with silver grains corresponding to light areas on the subject and clear areas with no silver grains corresponding to dark areas on the subject. In the second phase of wet-process photography, *printing,* the reversed light–dark image on the film is itself reversed on the *print.*

Print Paper.   Print paper is heavy white paper with an emulsion of silver halides on one surface. Unlike most films, print paper can be handled using a *safelight,* which is a weak light source covered with a colored filter that transmits light in a spectral region where the emulsion is not very sensitive. The print paper used in this project is Kodak *Polycontrast Rapid II RC Paper.* The term *polycontrast* indicates that different degrees of print contrast are obtained with different polycontrast (PC) filters used during printing. The PC filters are numbered in half steps from 1 through 4; the higher the number, the greater the contrast of the print. For this project, you can start with a PC 3 filter. The RC paper is available in two finishes, *glossy* (designated F) and *matte* (designated N). The glossy finish is used for research photomicrographs since journals require a glossy finish for publication.

Printing.   In the printing process, light is allowed to pass through the negative onto the print paper. Generally, we want to enlarge the image on the negative, a process requiring a *photographic enlarger* (Fig. 18.2). The main components of an enlarger are the lamp, filter drawer, condenser, negative carrier, and focusable lens system. When the negative is placed in the negative carrier and the enlarger lamp turned on, the image is projected onto the print paper below. The print paper is held flat in an *enlarger easel* placed on the baseboard. As light strikes the paper's emulsion, a latent image is formed. Just as with film, a developer is required to convert the latent image to a visible image, and fixer is required to remove the unexposed silver halide crystals, making the print permanent. In printing, it is necessary, once again, to compose the picture, focus the image, and determine the correct exposure.

Composition and Magnification.   When composing the print, you will have to obtain an appropriate degree of enlargement. The degree of enlargement is controlled with the height control knob (Fig. 18.2). The height may have to be adjusted several times until all the areas of interest appear in a projection of reasonable size. The areas of interest are framed using the enlarger easel; the easel can be moved around on the baseboard, and the movable borders of the easel can be adjusted. To obtain the exact magnification, the negative of the stage micrometer scale should be projected and later printed. Magnification is computed as follows:

$$\text{magnification} = \frac{\text{distance between scale divisions on the print}}{\text{distance between scale divisions on the stage micrometer}}$$

Focusing.   The focusing is done by adjusting the length of the enlarger bellows, which changes the distance between the negative and the lens. You will have to adjust the focus whenever the magnification is changed. The negative is in focus when the grains of the negative's emulsion appear in sharp focus on the print paper. For viewing the grains, you will need an *enlarger focusing scope* to magnify the image.

Determining the Correct Exposure.   Exposure, you will recall, is the product of light intensity and exposure time. In printing, light intensity is controlled by the

Lamphouse

Filter drawer

Condenser housing

Negative carrier

Bellows

Lens

Focusing knob

Height control knob

Baseboard

FIGURE 18.2
Photographic enlarger.
(Courtesy of Omega Division,
Berkey Marketing Companies.)

enlarger lens opening, expressed as the ratio of lens focal length/diameter of the lens aperture. This ratio is commonly referred to as the *f-stop*. Since the focal length of the lens is constant, a higher f-stop indicates a smaller lens aperture. In a typical series of f-stops ($f/4$, $f/5.6$, $f/8$, $f/11$, $f/16$, $f/22$), each higher f-stop lets through half as much light as does the preceding f-stop. Since $E = IT$, the same exposure would be obtained with 5 sec at $f/8$, with 10 sec at $f/11$, or with 20 sec at $f/16$.

There is a relatively simple method for determining the correct exposure for a particular negative. The procedure relies on a *projection print scale*, which is a sheet of celluloid divided into sectors. Each sector is in a different shade of gray and has a number (of seconds) printed on it. With the negative in the negative carrier, the appropriate PC filter in the filter drawer, the projection print scale overlying a fresh piece of print paper, and the enlarger lens aperture set at $f/8$, a 1-min exposure is made. The print is developed and the sector with the best exposure selected. The number in the sector with the best exposure is the approximate number of seconds (at $f/8$) required to obtain a print with that exposure. If the indicated exposure time is too short to gauge accurately, for example, 3 sec at $f/8$, you might want to expose the first print for 12 sec at $f/16$. A convenient range for exposure time is 10–15 sec. Should you want to print the same negative at a different magnification or at a different contrast, you will have to make a new exposure series with the projection print scale.

INSTANT-PROCESS METHODS

Black-and-White Prints.   A popular instant print film for photomicrography is Polaroid[1] *Type 667*. The film is very fast (ASA 3000) and, hence, rather grainy. With this pack film, a print is produced in about 1 min. The chemical processing begins automatically as the print is removed from the special camera back. Since there is no negative, print magnification is limited by the available microscope lenses, and print contrast can be controlled only with the condenser iris of the microscope and with colored filters.

Transparencies.   Black-and-white transparencies can also be processed quickly without a darkroom, using Polaroid *Polapan*[2] 35mm film. This film, which can be used in any 35mm camera, is a *reversal film*, meaning that it yields positive transparencies directly. Polapan is a continuous-tone, panchromatic, black-and-white film with high resolution, fine grain, and medium speed (ASA 80).

After the film has been exposed and rewound, the cartridge with its film leader is placed in the *Polaroid 35mm AutoProcessor*, which is a "black box" with gears, rollers, lever, and crank. Also placed in the processor is a *processing pack*, which contains the developing chemicals. Turning the crank brings the film and developing chemicals into contact. Development is allowed to proceed for 60 sec, after which the film is rewound and removed from the processor. The film is then completely dry and can immediately be cut, mounted, and projected.

PROCEDURES

PHOTOGRAPHING CELL PREPARATIONS WITH TECHNICAL PAN FILM

1. Load the film in the 35mm camera back and advance the film three frames. After the camera is loaded and mounted, pull out the dark slide.
2. Place the microscope slide on the stage and adjust the microscope for Köhler illumination (see Project 1, p. 7). After you have located the field you want to photograph, adjust the condenser iris for optimum contrast and insert a No. 58 or No. 11 filter.
3. While looking through the focusing eyepiece, position the microscope slide so that the structures you would like to photograph are included in the correct reticle frame. Turn the eyepiece to include all the structures of interest and to obtain the best composition.
4. Turn the camera back so that it is at the same angle as the reticle frame.
5. Turn the collar of the focusing eyepiece until the cross hairs on the reticle are in sharp focus. Then, while looking through the focusing eyepiece, bring the object into sharp focus with the fine adjustment knob. If eyeglasses are routinely worn, they should be left on for the entire focusing procedure. At this point, depending on the design of your photomicrographic assembly, you may have to divert the light from the focusing eyepiece to the camera.
6. Set the exposure meter for the film's ASA speed (50) and, if required, for the film format (35mm) and ocular magnification. Record the meter reading and all other information about the photomicrograph on Data Sheet 18.1.
7. Make an exposure series, bracketing the indicated exposure time. Each successive film frame should have twice the exposure of the preceding one, with the middle frame of the series having the indicated exposure time. So, for

[1]Polaroid is a trademark of Polaroid Corporation, Cambridge, Mass.
[2]Polapan is a trademark of Polaroid Corporation, Cambridge, Mass.

PHOTOMICROGRAPHY

example, if the exposure meter indicates 1/15 sec, a series of three frames could be used with exposure times of 1/30, 1/15, and 1/8 sec. Be sure to advance the film after each exposure and to record the exposure time for each frame on Data Sheet 18.1.

8. Photograph the stage micrometer scale so that you will be able to determine the exact magnification of the print.

PROCESSING TECHNICAL PAN FILM

Photographic chemicals should be handled with care; they are toxic if swallowed or inhaled, and repeated contact may cause skin irritation and allergic skin reaction. Always wipe up spills immediately, as neatness and cleanliness are essential to quality photomicrographs. The developer should be at 20° C and the other solutions and rinses at 18.5°–21° C. The diluted developer is discarded after use, but the stop bath and fixer can be reused.

1. Rewind the exposed film and remove the magazine from the camera.
2. In *total darkness*, and handling the film only by the edges and leader, load the exposed film on the developing reel. Place the loaded reel in the developing tank and then screw on the cover. The closed developing tank has openings in the cover for adding and pouring out liquids, but is lightproof, so that the processing can be done with the room lights on.
3. Make a 1:20 dilution of the *HC-110* developer stock solution and, using a water bath, adjust the temperature of the diluted stock solution to 20° C. As 300 ml will usually cover one developing reel, mix 15 ml of the HC-110 stock solution with 285 ml of water in a 600-ml beaker.
4. Add the diluted HC-110 stock solution to the developing tank and immediately rotate the reel back and forth several times with the agitator rod to dislodge air bubbles. Develop for 8 min, agitating the solution by rotating the reel for 5 sec every 30 sec, first clockwise and then counterclockwise. After 8 min, discard the developer by inverting the closed developing tank over the sink drain.
5. Immediately add about 300 ml of stop bath to the developing tank. Agitate for 15–30 sec and then return the stop bath to its original container.
6. Add about 300 ml of fixer to the developing tank. Fix for 2–4 min with frequent agitation and then return the fixer to its original container.
7. Open the developing tank and wash the film under running tap water for 5–15 min. The stream of water should run into the center of the reel within the tank. Wash time is reduced to only 1 min (and water conserved) if the film is first soaked in a solution of Kodak *Hypo Clearing Agent*.
8. Pour off the water and add about 300 ml of a solution of Kodak *Photo Flo* to the developing tank. A 30-sec soak in this wetting solution will prevent the formation of water spots on the film as it dries.
9. Remove the reel from the tank, separate the two parts of the reel, attach film clips to the ends of the film, and hang the film in a dust-free area to dry. When completely dry, cut the film into lengths of five or six frames and store in glassine envelopes.

PRINTING

Photographic chemicals should be handled with care; they are toxic if swallowed or inhaled, and repeated contact may cause skin irritation and allergic skin reaction. Always wipe up spills immediately, as neatness and cleanliness are essential to quality photomicrographs.

The developer should be at 20° C and the other solutions and rinses at 18.5°–24° C.

The diluted developer, stop bath, and fixer can be used for multiple prints during the same printing session. Always make certain that before the room light or enlarger lamp is turned on, unused print paper is in its original package or in a lightproof cabinet.

1. Turn on the safelight. The print paper can be handled with an OC safelight filter in a lamp with a 15-watt bulb kept at least 1.2 m (4 feet) from the paper. The print paper should not be exposed to the safelight (at 1.2 m) for more than 3 min.

2. With the room lights on, add the diluted developer (1 vol *Dektol* + 2 vol water), stop bath, and fixer to three trays.

3. Select the best negative in the exposure series (see p. 235).

4. Lightly brush any dust off the negative, using a camel's hair brush. Then place the negative, emulsion side down, in the negative carrier of the enlarger.

5. Turn off the room lights and place a piece of scrap print paper in the enlarger easel, emulsion side up.

6. Turn on the enlarger lamp and open the lens aperture all the way (lowest *f*-stop).

7. Using the height control knob, adjust the enlarger height above the easel until you obtain the approximate degree of enlargement desired. You will have to adjust the focus each time the enlarger height is changed. This is done by changing the length of the bellows. Compose the print by moving the entire easel and its borders. When you are satisfied with the composition, make the final adjustment in focus. Use an enlarger focusing scope to see when the grains of the negative's emulsion are in sharp focus on the scrap paper. When the image is in focus, set the lens aperture to *f*/8 and turn off the enlarger lamp.

8. Insert a PC 3 filter in the filter drawer of the enlarger.

9. Place a fresh piece of print paper (cut to size) in the easel and place a projection print scale on top of the paper.

10. Turn on the enlarger lamp and expose the print paper for 1 min.

11. Develop the exposed print paper by immersing it, emulsion side down, into the developer and then immediately turning it over. Develop for 1 min with continuous agitation. The print should be handled with tongs, *not* fingers.

12. Drain the print by touching it to the upper edge of the tray and then transfer it to the stop bath. Agitate the print in the stop bath for 5 sec.

13. Drain and transfer the print to the fixer. Fix for 2 min with frequent agitation. After fixing, you can turn on the room lights.

14. Wash the print in running tap water for 4 min.

15. Examine the print and select the sector with the best exposure. The number printed in this sector is the approximate number of seconds (at *f*/8) required to obtain a print with that exposure. For the final print, you will want to use an exposure time between 10 and 15 sec. The lens aperture may have to be opened wider or stopped down, and the exposure time changed accordingly.

16. Make the final print, as earlier, but with the exposure time (and appropriate *f*-stop) that you have determined to be best. It is a common practice to make several prints, using a slightly different exposure time for each. An efficient way to make two or three print exposures is to make *all* the exposures first and then develop the several prints at the same time. When making each exposure, be sure that the other pieces of print paper are in a lightproof cabinet. When developing the several pieces of exposed print paper, continually interleaf them in the developing tray so that air bubbles will not be trapped next to the emulsion. To determine the exact magnification of the photomicrograph, print the negative of the stage micrometer scale. If you want to make prints with higher or lower contrast, replace the PC 3 filter with a PC 4 or PC 2 filter, respectively, and make a new test exposure using the projection print scale. Be sure that the enlarger lamp and safelight are turned off when you have finished.

17. After the prints have been washed, wipe the surface with a squeegee and allow to air dry at room temperature. The dry prints should be trimmed with a paper cutter and mounted on lightweight cardboard with *mounting tissue*, which requires a photographic tacking iron to apply.

PHOTOGRAPHING CELL PREPARATIONS WITH POLAROID TYPE 667 FILM

The Polaroid developing process uses a caustic jelly, which is spread between the print surface and the black backing. If any should accidentally get on your skin, immediately wipe it off and wash the area with plenty of water. Keep the jelly away from eyes, mouth, and clothing.

1. Load the film pack in the Polaroid camera back, following the instructions that accompany the film. Pull out the black paper shield.

2. Place the microscope slide on the stage and adjust the microscope for Köhler illumination (see Project 1, p. 7). After you have located the field you want to photograph, adjust the condenser iris for optimum contrast and insert a No. 58 or No. 11 filter in the light path.

3. While looking through the focusing eyepiece, position the microscope slide so that the structures you would like to photograph are included in the correct reticle frame. Turn the eyepiece to include all the structures of interest and to obtain the best composition.

4. Turn the camera back so that it is at the same angle as the reticle frame.

5. Turn the collar of the focusing eyepiece until the cross hairs on the reticle are in sharp focus. Then, while looking through the focusing eyepiece, bring the object into sharp focus with the fine adjustment knob. If eyeglasses are routinely worn, they should be left on for the entire focusing procedure. At this point, depending on the design of your photomicrographic assembly, you may have to divert the light from the focusing eyepiece to the camera.

6. Set the exposure meter for the film's ASA speed (3000) and, if required, for the film format (3¼ in. × 4¼ in.) and ocular magnification. Record the meter reading and all other information about the photomicrograph on Data Sheet 18.1.

7. Make an exposure series. If there is a dark slide close to the film plane of the camera, a total of five exposures can be made on the same print. First, pull out the dark slide as far as it will go. Expose the film for one-fourth of the indicated exposure time and then push in the dark slide about one-fifth of the way. Again, expose the film for one-fourth of the indicated exposure time and push in the dark slide another one-fifth of its travel. Repeat this procedure with one-half of the indicated exposure time, with the indicated exposure time, and with twice the indicated exposure time. Had the meter reading been 1/15 sec, the one print would now have five sections with exposure times of 1/60, 1/30, 1/15, ~1/8, and ~1/4 sec. The total exposure time in any section is the sum of all the exposure times received prior to covering that section with the dark slide. Make a note of the exposure times.

8. While holding the camera back securely, pull the white tab out of the camera, after which a yellow tab will appear. Immediately (again, holding the camera securely), pull the yellow tab *straight* out of the camera, using a steady, uninterrupted pull at moderate speed. After it is all the way out, note the time, as development has begun. After the recommended development time, which is 45 sec at 21° C (70° F) and 30 sec at 24° C (75° F) or above, quickly separate the print from the black backing, starting at a corner of the print at the end nearer the yellow tab. Fold the black backing, moist side in, and discard it.

9. Examine the print and select the section with the best exposure (the longer the exposure, the lighter the print). Then expose the next sheet of film, using the best exposure time. Record the exposure time on Data Sheet 18.1.

PHOTOGRAPHING CELL PREPARATIONS WITH POLAPAN FILM

1. Load the film in the 35mm camera back and advance the film three frames. After the camera is loaded and mounted, pull out the dark slide.

2. Place the microscope slide on the stage and adjust the microscope for Köhler illumination (see Project 1, p. 7). After you have located the field you want to photograph, adjust the condenser iris for optimum contrast and insert a No. 58 or No. 11 filter.

3. While looking through the focusing eyepiece, position the microscope slide so that the structures you would like to photograph are included in the correct reticle frame. Turn the eyepiece to include all the structures of interest and to obtain the best composition.

4. Turn the camera back so that it is at the same angle as the reticle frame.

5. Turn the collar of the focusing eyepiece until the cross hairs on the reticle are in sharp focus. Then, while looking through the focusing eyepiece, bring the object into sharp focus with the fine adjustment knob. If eyeglasses are routinely worn, they should be left on for the entire focusing procedure. At this point, depending on the design of your photomicrographic assembly, you may have to divert the light from the focusing eyepiece to the camera.

6. Set the exposure meter for the film's ASA speed (80) and, if required, for the film format (35mm) and ocular magnification. Record the meter reading and all other information about the photomicrograph on Data Sheet 18.1.

7. Make an exposure series, bracketing the indicated exposure time. Each successive film frame should have twice the exposure of the preceding one, with the middle frame of the series having the indicated exposure time. So, for example, if the exposure meter indicates 1/15 sec, a series of three frames could be used with exposure times of 1/30, 1/15, and 1/8 sec. Be sure to advance the film after each exposure and to record the exposure time for each frame on Data Sheet 18.1.

PROCESSING POLAPAN FILM

The processing pack contains a caustic jelly. If any should accidentally leak out, keep it away from eyes, mouth, and clothing. If any gets on your skin, immediately wipe it off and wash the area with plenty of water. Immediately after development, the processing pack should be discarded in the original box.

1. Rewind the exposed film and remove the cartridge from the camera.
2. Load the film and processing pack into the AutoProcessor, following the instructions that accompany the unit.
3. Close the cover of the AutoProcessor, making certain that it latches.
4. Grasp the lever at the side of the processor and flip it down at a steady, medium rate of speed.
5. Wait 5 sec and then turn the crank *fast*, at about *two* revolutions per second, in a *clockwise* direction. (Turning it counterclockwise will damage the unit.) As soon as the gear clicking is no longer heard, immediately stop turning the crank.
6. Wait 60 sec. This is the correct processing time for this film at 21° C (70° F). At lower temperatures, the processing time must be increased (see film instructions).

7. Grasp the lever and flip it up at a steady, medium rate of speed.

8. Turn the crank *very fast*, at about *three* revolutions per second, in a clockwise direction (*not counterclockwise*). When you no longer hear the gear clicking, continue turning the crank for an additional three or four revolutions.

9. Open the processor cover. Remove and discard the processing pack and then remove the film cartridge. The film is dry and can immediately be cut and mounted using the Polaroid *35mm Slide Mounter* and Polaroid *35mm Slide Mounts*. The mounted transparencies are then ready for projection.

REFERENCES

Basic Developing, Printing, Enlarging in Black-and-White. 1977. Kodak Publication No. AJ-2, Eastman Kodak Co., Rochester, N.Y.

Lefkowitz, L. 1985. *Polaroid 35mm Instant Slide System: A User's Manual.* Focal Press, Boston.

Loveland, R. P. 1970. *Photomicrography: A Complete Treatise*, Vols. 1 and 2. John Wiley, New York.

Photography Through the Microscope, 7th ed. 1980. Kodak Publication No. P-2. Eastman Kodak Co., Rochester, N.Y.

Data Sheet 18.1
Checklist of important variables in photomicrography.

Film, ASA _____

Frame	Slide no., stage coord.	Tissue, stain	Trans. setting	Objective, ocular	Filters	Meter reading	Exp. time	Other info.
1								
2								
3								
4								
5								
6								
7								
8								
9								
10								
11								
12								
13								
14								
15								
16								
17								
18								
19								
20								

INSTRUCTOR'S GUIDE

This section lists the equipment, supplies, and reagents required for each project. Quantities are calculated for a class of 20 students working individually or in teams, as noted.

Note to the preparator: Most of the required reagents are toxic to some degree. They should never be pipetted by mouth, and any spills on skin or clothing should be washed off immediately with water.

PROJECT 1

Equipment

20 compound microscopes, each equipped with an ocular micrometer, green filter, and 10×, 40×, and 100× objectives

Phase-contrast microscopes (optional)

Supplies

Micrometer slides	1 onion
Microscope slides	3 droppers
Coverslips, 22 mm², No. 1½	10 pairs pointed forceps
10 single-edge razor blades	Toothpicks
5 fresh sprigs of *Elodea*	Kimwipes

1 culture of *Euglena gracilis*. The culture should last for at least 1–2 weeks if kept in a well-illuminated area (but out of direct sunlight) with the temperature between 18°–22° C. For long-term maintenance of the culture, use split pea medium: boil 40 yellow split peas in 1 liter of distilled water for 10 min. Allow the medium to cool and then inoculate immediately.

20 prepared slides of *Zea* stem, cross section

20 prepared slides of common bacterial forms (cocci, bacilli, spirilla)

Lens paper, enough for the entire semester

Reagents

100 ml distilled water dispensed in five dropping bottles

100 ml Lugol's iodine solution dispensed in five dropping bottles: dissolve 6.0 g KI and 4.0 g I in 100 ml distilled water.

5 bottles of Protoslo (Carolina Biological Supply Co.)

Immersion oil, enough for the entire semester

PROJECT **2**

Each student should be provided with two unstained, fixed blood smears, prepared any time up to 6 hr before class. The blood should come from a healthy individual other than a student. Care should be taken when handling blood since there is always the slight possibility that the blood is contaminated with infectious agents. Accordingly, disposable gloves should be worn, spills wiped up, the area disinfected, and waste materials disposed of properly. It is advisable that the individual make the smears using his/her own blood. For making the fixed blood smears, follow this procedure:

1. Clean the slides thoroughly. New slides from the box should always be degreased by dipping in 95% ethanol and wiping with a lint-free tissue, like Kimwipes.
2. Use aseptic technique to obtain the blood sample. After washing the hands with soap and water, and drying them, clean the tip of the ring or middle finger with a sterile alcohol swab. When the fingertip has dried, lance it with a sterile lancet, squeeze the finger, and wipe away the first drop of blood. *The lancet should never be used more than once and is to be discarded immediately.*
3. Again, squeeze the finger and place a drop of blood near the frosted end of one slide. Hold a second slide with its edge at about a 30° angle on the first and bring it toward the drop; see (a) in the accompanying figure. After contact is made and the drop spreads along the edge of the slide (b), push the second slide to the other end in one smooth motion (c). In the preparation of multiple smears, the slide used to push should be the next one on which the smear is made. After the smears have been made, clean the lanced fingertip with alcohol.
4. Allow the smears to air dry.
5. Fix the air-dried smears in a 9:1 mixture of absolute ethanol:formalin for 10 min, in a Coplin jar.
6. Pour off and discard the fixative, using forceps, not fingers, to restrain the slides.
7. Rinse the slides in the Coplin jar under running tap water for 2 min. Then replace with distilled water and pour it off.
8. Remove the slides from the Coplin jar and allow them to air dry.

The steps (a), (b), and (c) for preparing a thin blood smear. See text for explanation.

As a check on whether the buffer alone removes PAS-positive material, one or several students can place the slide without diastase treatment in buffer at 20° C. As a check on whether preexisting aldehydes are being stained by Schiff's reagent, one or several students can be assigned an additional control slide that is not subjected to periodic acid treatment.

Equipment
20 compound microscopes

Supplies

Microscope slides with frosted end	Sterile alcohol swabs
Coverslips, 22 × 40 mm	Sterile blood lancets
2 10-ml graduated cylinders	Parafilm
2 250-ml graduated cylinders	Kimwipes
20 pairs forceps	Benchkote

Slide holders to prop slides in a near vertical position. Available from Thomas Scientific.

20 Wright- or Giemsa-stained human blood smears. They can be purchased from Carolina Biological Supply Co. or Ward's Natural Science Establishment, Inc. Alternatively, they can be prepared simply and quickly using VWR Stat Stain (available from VWR Scientific).

Reagents

40 ml 95% ethanol in a Coplin jar

200 ml 9:1 ethanol:formalin fixative dispensed in five Coplin jars: mix 180 ml absolute ethanol and 20 ml formalin (37% w/w formaldehyde). Prepare shortly before use.

120 ml 1% diastase solution dispensed in three Coplin jars: dissolve 1.20 g diastase of malt (Fisher Scientific) in 120 ml M/15 phosphate buffer, pH 6.0. Buffer: dissolve 3.81 g KH_2PO_4, 0.76 g Na_2HPO_4 in distilled water to make 500 ml of solution. The diastase solution is good for 1 week if refrigerated.

300 ml 1% periodic acid dispensed in eight Coplin jars: dissolve 3.0 g H_5IO_6 in 300 ml distilled water. The solution is good for several weeks if stored in a dark bottle.

100 ml 1 N HCl: in a fume hood, dilute 8.1 ml concentrated HCl to 100 ml with distilled water.

220 ml Schiff's reagent (method of C. Leuchtenberger) dispensed in seven Coplin jars: dissolve 1.0 g basic fuchsin in 200 ml boiling distilled water, with boiling chips, in a 500-ml Erlenmyer flask. Boil for 5 min with continual agitation. Cool to 50° C, filter with suction through Whatman No. 1 filter paper, and add 20 ml 1 N HCl. Cool to 25° C, add 2.0 g Na or K metabisulfite, and shake well. Allow the solution to remain in a stoppered bottle in the dark for 16–24 hr. Add 0.5 g activated charcoal (Norit, neutral, available from Fisher Scientific), shake well for 1 min, and filter with suction through two pieces of coarse filter paper. The solution should be colorless or nearly so; if it is still yellow, add more charcoal and refilter. Store at 4° C in a dark stoppered bottle. Schiff's reagent can be reused and is good as long as it remains colorless or only a light-straw color.

100 ml 10% metabisulfite solution: dissolve 10 g Na or K metabisulfite in distilled water to make 100 ml of solution.

75 ml Harris's hematoxylin (any commercial brand) dispensed in five Coplin jars. Shake well before dispensing.

175 ml xylene dispensed in five Coplin jars

50 ml Permount dispensed in five glass-rod dropping bottles

Each student should be provided with two unstained, fixed blood smears, prepared any time up to 6 hr before class. The blood should come from a healthy individual other than a student. Care should be taken when handling blood since there is always the slight possibility that the blood is contaminated with infectious agents. Accordingly, disposable gloves should be worn, spills wiped up, the area disinfected, and waste materials disposed of properly. It is advisable that the individual make the smears using his/her own blood. For making the fixed blood smears, follow the procedure on page 248, except that here (1) use a larger drop of blood so that the smear will be thicker and (2) use Carnoy's fixative, a 6:3:1 mixture of absolute ethanol:chloroform:glacial acetic acid.

As a check on whether the distilled water alone removes RNA, one or several students can place the slide without RNase treatment in distilled water at 37° C.

Equipment

20 compound microscopes   1 37° C water bath

Supplies

Microscope slides with frosted end   Sterile alcohol swabs
Coverslips, 22 × 40 mm   Sterile blood lancets
20 pairs forceps   Kimwipes
Benchkote

Reagents

40 ml 95% ethanol in a Coplin jar

200 ml Carnoy's fixative dispensed in five Coplin jars: mix 120 ml absolute ethanol, 60 ml chloroform, and 20 ml glacial acetic acid. Prepare shortly before use.

120 ml 0.1% RNase solution dispensed in three Coplin jars: dissolve 120 mg RNase (Sigma Chemical Co., cat. no. R4875) in 120 ml distilled water. Adjust the pH to 6.5–7.0 with 0.1 N NaOH. Solution is good for 1 week if refrigerated.

175 ml methyl green–pyronin staining solution (method of E. B. Taft) dispensed in five Coplin jars: dissolve 1.0 g of the *pure dye* methyl green (Harleco) in 200 ml hot distilled water. For example, if a batch of methyl green has a dye content given as 89% on the label, add 1.12 g. When cool, transfer to a separatory funnel and extract repeatedly with 30-ml aliquots of chloroform until the chloroform is colorless or nearly so. The chloroform removes the contaminant, methyl violet. Add 0.1 g of the *pure dye* pyronin Y (Kodak Laboratory Chemicals) to 200 ml of the chloroform-extracted methyl green solution. For example, with a batch of pyronin Y that has a dye content of 50% and with 190 ml of chloroform-extracted methyl green solution, add 0.19 g. Shake well and dispense. The staining solution, which should be stored in a dark stoppered bottle, can be reused and is good for one week.

200 ml 3:1 *tert*-butanol:ethanol dispensed in five Coplin jars: mix 150 ml *tert*-butanol and 50 ml absolute ethanol.

350 ml xylene dispensed in 10 Coplin jars

50 ml Permount dispensed in five glass-rod dropping bottles

Students should work in pairs; quantities are calculated for 20 students working in 10 teams. The homogenization should be done by one designated team. To confirm that the absorbance of DNA with the orcinol reaction is about 10% the absorbance of the same concentration of RNA, one or several teams can run the orcinol reaction with a DNA stock solution.

Equipment

UV spectrophotometer(s)

2 clinical centrifuges

1 Waring blender, jar thoroughly chilled

2 90° C water baths

10 hot plates

1 balance

10 spectrophotometers. *Note*: For measurements above 600 nm, the B & L Spectronic 20 should be equipped either with the wide-range (400–700 nm) or red-sensitive (600–950 nm) phototube and filter.

Supplies

120 glass cuvettes

30 quartz cuvettes

30 g calf liver, frozen

10 15-ml centrifuge tubes

10 10-ml graduated cylinders

50 1-ml pipets

60 2-ml pipets

20 5-ml pipets

10 pipet fillers, rubber bulb type

130 test tubes, 16 × 125 mm, with screw caps

Pasteur pipets, 146 mm, with dropper bulbs

20 400-ml beakers for boiling-water and ice-water baths

10 test rube racks

10 permanent marking pens

Parafilm

Kimwipes

20 pairs protective goggles

Boiling chips

10 French curve templates and rulers

10 10-ml pipets

Reagents

100 ml 0.15 M NaCl–0.015 M Na$_3$ citrate (saline–sodium citrate or SSC): dissolve 4.38 g NaCl, 2.20 g Na$_3$ citrate · 2H$_2$O in distilled water to make 500 ml of solution. Adjust the pH to 7.0 with 1 N HCl.

100 ml DNA stock solution, 100 µg/ml in SSC: dissolve 10 mg DNA (e.g., calf thymus DNA, Sigma Chemical Co., cat. no. D1501) in 100 ml of the saline–sodium citrate solution at room temperature. Check the absorbance at 260 nm and, if necessary, adjust the DNA concentration so that $A \cong 1$.

100 ml RNA stock solution, 50 µg/ml in SSC: dissolve 5 mg RNA (e.g., calf liver RNA, Sigma Chemical Co., cat. no. R7250) in 100 ml of the saline–sodium citrate solution at room temperature. Check the absorbance at 260 nm and, if necessary, adjust the RNA concentration so that $A \cong 1$.

120 ml ice-cold distilled water

150 ml ice-cold 10% (w/v) trichloroacetic acid (TCA) dispensed in 10 capped Erlenmyer flasks

350 ml 5% (w/v) TCA dispensed in 10 capped Erlenmyer flasks

250 ml 95% ethanol dispensed in 10 capped Erlenmyer flasks

100 ml each of four DNA stock solutions: 400 µg/ml, 300 µg/ml, 200 µg/ml, 100 µg/ml, dispensed in capped test tubes. To make the 400 µg/ml stock solution, add 100 mg DNA (e.g., calf thymus DNA, Sigma Chemical Co., cat. no. D1501) to 250 ml of 5% TCA. To dissolve, place the mixture in a 90° C water bath for several minutes, swirling occasionally. To make the 300 µg/ml stock solution, dilute 75 ml of the 400 µg/ml stock solution to 100 ml with 5% TCA. To make the 200 µg/ml stock solution, dilute 50 ml of the 400 µg/ml stock solution to 100 ml with 5% TCA. To make the 100 µg/ml stock solution, dilute 25 ml of the 400 µg/ml stock solution to 100 ml with 5% TCA.

100 ml each of four RNA stock solutions: 400 µg/ml, 300 µg/ml, 200 µg/ml, 100 µg/ml, dispensed in capped test tubes. Follow the same procedure used for making the four DNA stock solutions but, instead of DNA, use calf liver RNA (Sigma Chemical Co., cat. no. R7250).

300 ml diphenylamine reagent dispensed in 50-ml burets: in a fume hood, dissolve 3.0 g fresh or recrystallized (in petroleum ether) diphenylamine in 300 ml glacial acetic acid and 8.25 ml concentrated sulfuric acid. Prepare within 30 min before use.

250 ml orcinol reagent dispensed in 50-ml burets: in a fume hood, dissolve 2.5 g orcinol in 250 ml concentrated HCl containing 1.25 g $FeCl_3 \cdot 6H_2O$. Prepare within 30 min before use.

PROJECT **5**

Students should work in teams of four; quantities are calculated for 20 students working in five teams. Each electrophoresis chamber accommodates up to three cellulose acetate plates, so that a total of two electrophoresis chambers and one power supply should be ample for an entire class.

The hemoglobin samples should be handled with care since there is always the slight possibility that the blood is contaminated with infectious agents. Accordingly, disposable gloves should be worn by students and others handling the hemoglobin samples, spills wiped up, the area disinfected, and waste materials disposed of properly. *The electrophoresis equipment should always be supervised because of the potential shock hazard associated with the power supply.*

The procedures assume the use of Titan III-H Cellulose Acetate Plates, the Super Z Applicator Kit, and Titan Gel Chamber, available from Helena Laboratories, 1530 Lindbergh Drive, P.O. Box 752, Beaumont, Texas 77704-0752. The trade names of all Helena electrophoresis items (including the registered trademarks *Titan, Supre Heme,* and *Zip Zone*) and catalog numbers are also listed. The project can also be carried out using equipment and supplies available from Gelman Sciences Inc., 600 South Wagner Road, Ann Arbor, Michigan 48106.

Equipment

2 Titan Gel Chambers (4063)   1   56° C oven
1 Titan Power Supply, 110 volts
   (1500)

Supplies

5 Drummond Dialamatic Microdispensers (micropipetters), 10 µl (6210)

1 box 25 Titan III-H Cellulose Acetate Plates, 60 × 76 mm (3022)

5 Super Z Applicator Kits, 8 sample (4088). Each kit contains 1 eight-sample applicator, 2 eight-sample well plates, 1 aligning base, 1 bottle Zip Zone Prep detergent.

1 Bufferizer (containers for soaking plates) (5093)

1 700 Staining Set (5114). Contains 1 carrying rack for plates and 7 700-ml plastic containers with lids.

1 box 100 Titan Blotter Pads, 89 × 108 mm (5037)

1 box 500 Disposable Wicks (5081)

1 pkg 200 Titan Plastic Envelopes, 121 × 64 mm (5052)

Disposable gloves

1 100-ml graduated cylinder

5 permanent marking pens

Benchkote

Reagents

1 box of 10 pkg Supre Heme Buffer (5082). Dissolve the contents of one package in distilled water to make 1000 ml of solution. The buffer in the electrophoresis chamber should be discarded after use. Buffer used to soak the cellulose acetate plates can be reused for soaking up to 12 plates.

AA$_2$ Hemo Control (5328)

ASA$_2$ Hemo Control (5329)

AFSC Hemo Control (5331)

5 250-ml bottles Ponceau S Stain (5525)

2 liters 5% acetic acid: bring 100 ml glacial acetic acid to 2 liters with distilled water.

PROJECT 6

Students should work in teams of four; quantities are calculated for 20 students working in five teams. Each team is responsible for obtaining one of the tissue extracts and then dispensing the diluted extract to other members of the class. Each team can then electrophorese extracts from four different tissues. As an alternative to using extracts prepared in class, commercially prepared tissue extracts from rat heart and liver can be used (LD Isozyme Control, cat. no. 5919, Helena Laboratories). Each vial of lyophilized extract is reconstituted with 1.0 ml distilled water. The reconstituted extract is applied (3.0 μl) to the agarose gel plate *without dilution*. If this option is selected, be sure to follow all safety precautions and recommendations for storage given on the package insert.

Each electrophoresis chamber accommodates two agarose gel plates, so that a total of three electrophoresis chambers and two power supplies should be ample for an entire class. *The electrophoresis equipment should always be supervised because of the potential shock hazard associated with the power supply.*

The procedures assume the use of the Titan Gel Isoenzyme Kit and Titan Gel Chamber, available from Helena Laboratories, 1530 Lindbergh Drive, P.O. Box 752, Beaumont, Texas 77704-0752. The trade names of all the Helena electrophoresis items (including the registered trademark *Titan*) and catalog numbers are also listed.

Electrophoresis Equipment

3 Titan Gel Chambers (4063)   1 45° C incubator

2 Titan Power Supplies, 110 volts 1 60° C oven
(1500)

1 densitometer (optional)

Electrophoresis Supplies

5 Drummond Dialamatic Microdispensers (micropipetters), 10 μl (6210)

1 Titan Gel Isoenzyme Kit (3043). Contains 10 agarose gel plates, 10 vials LDH isoenzyme reagent, diluent, buffer, templates, and blotters A, B, and D. The items supplied with the kit may change occasionally. Minor changes in Helena Laboratories' recommended procedures may occur; instructions are supplied with each kit.

5 Titan Gel Isoenzyme Incubation Chambers (4062)

1 box development slides (76 × 102 mm glass plates) (5008)

5 development weights (5014)

1 pkg 100 Titan Plastic Envelopes, 120 × 102 mm (5053)

5 5-ml pipets

5 large finger bowls, 15–20 cm in diameter

1 50-ml graduated cylinder

5 permanent marking pens

Electrophoresis Reagents

LDH buffer, pH 8.1-8.3 (supplied with Titan Gel Isoenzyme Kit). Dissolve the contents of the package in distilled water to make 1500 ml of solution. The buffer in the electrophoresis chamber should be discarded after use.

LDH isoenzyme reagent (NAD, lithium lactate, nitro blue tetrazolium, phenazine methosulfate; supplied with Titan Gel Isoenzyme Kit). Reconstitute the contents of each vial with 1.0 ml diluent (supplied) about 5 min before electrophoresis is completed. Invert each vial repeatedly until all material is dissolved. All the reconstituted reagent in each vial is used for one gel plate.

2 liters 10% acetic acid: bring 200 ml glacial acetic acid to 2 liters with distilled water.

Homogenization Equipment

5 10-ml Potter–Elvehjem tissue grinders

2–4 stirrer motors (e.g., Wheaton overhead stirrer)

1 balance

1 refrigerated centrifuge or a clinical centrifuge, prechilled and run in a cold room or refrigerator

Homogenization Supplies

1 rat	5 pipet fillers, rubber bulb type
1 dissecting tray	10 centrifuge tubes
Disposable gloves	20 5-ml vials
5 wooden chopping blocks	2 50-ml graduated cylinders
5 single-edge razor blades	5 50-ml Erlenmyer flasks
5 1-ml pipets	5 400-ml beakers for ice-water baths
5 2-ml pipets	Parafilm

Dissecting instruments: scalpel, scissors, forceps, dissecting pins

Homogenization Reagents

2 liters ice-cold distilled water

PROJECT **7**

Equipment

20 compound microscopes equipped with ocular micrometers

Supplies

Microscope slides with frosted end	20 pairs forceps
Coverslips, 22 mm^2, No. 1½	Kimwipes
10 fresh sprigs of *Elodea*	

Reagents

100 ml of each of the following sucrose solutions dispensed in five dropping bottles. Add the indicated amount of sucrose to a volumetric flask and bring the volume to 100 ml with distilled water: 0.2 M sucrose (6.85 g), 0.3 M sucrose (10.27 g), 0.4 M sucrose (13.69 g), 0.5 M sucrose (17.12 g), 0.6 M sucrose (20.54 g). Store at 4° C. Just before use at room temperature, swirl each dropping bottle.

100 ml of each of the following alcohol-sucrose solutions dispensed in five dropping bottles. Add the indicated volume of the alcohol to a volumetric flask and bring the volume to 100 ml with 0.3 M sucrose: 0.4 M methanol–0.3 M sucrose (1.6 ml), 0.4 M ethanol–0.3 M sucrose (2.3 ml), 0.4 M ethylene glycol–0.3 M sucrose (2.2 ml), 0.4 M 1-propanol–0.3 M sucrose (3.0 ml), 0.4 M propylene glycol–0.3 M sucrose (2.9 ml), 0.4 M glycerol–0.3 M sucrose (2.9 ml). Store at 4° C. Just before use at room temperature, swirl each dropping bottle.

PROJECT **8**

For the cell fractionation procedures, students should work in teams of four; quantities are calculated for 20 students working in five teams. For the microscopic examination of the fractions, students should work individually. The bulk of the project can be completed even if a refrigerated, high-speed centrifuge is not available since both the chloroplast fraction of spinach and the nuclear fraction of cauliflower can be isolated with a clinical centrifuge. For the nuclear fraction, the centrifuge should be prechilled and run in a cold room or refrigerator.

Equipment

    1 clinical centrifuge with a swinging-bucket rotor
    1 centrifuge tachometer
    1 refrigerated, high-speed centrifuge
  20 compound microscopes equipped with ocular micrometers

Supplies

10	mortars and pestles, chilled	Microscope slides
5	10-ml graduated cylinders	Coverslips, 22 mm^2
5	25-ml graduated cylinders	10 funnels
5	50-ml graduated cylinders	5 spatulas
5	single-edge razor blades	5 pairs scissors
5	400-ml beakers for ice-water baths	Parafilm
5	15-ml test tubes	Kimwipes
5	small test tubes, 10 × 75 mm	5 rulers
12	15-ml round bottom centrifuge tubes	

Pasteur pipets, 146 mm, and dropper bulbs

200 g fresh spinach. Just before use, rinse with cold tap water and blot dry.

  2 fresh, firm heads of cauliflower, refrigerated until used. Just before use, rinse with cold tap water and blot dry.

  6 50-ml centrifuge tubes for first centrifugation (for obtaining the nuclear fraction with a refrigerated, high-speed centrifuge). With a clinical centrifuge, 15-ml conical centrifuge tubes can be used.

  6 50-ml or 10 25-ml centrifuge tubes for second centrifugation (for obtaining the mitochondrial fraction)

Cotton cheesecloth, grade 10 (Curity). Available from Thomas Scientific.

Reagents

50 g purified sea sand

250 ml 0.35 M NaCl–0.02 M Tris buffer, pH 7.5: dissolve 5.11 g NaCl, 0.61 g tris (hydroxymethyl) aminomethane in distilled water to make 250 ml of solution. Adjust pH to 7.5 with 6 N HCl. Store at 4° C.

100 ml of a saturated solution of urea in distilled water, dispensed in five dropping bottles

100 ml 2% (v/v) Triton X-100 in distilled water, dispensed in five dropping bottles

500 ml mannitol grinding medium, 0.3 M D-mannitol-0.02 M phosphate buffer, pH 7.2: dissolve 27.33 g D-mannitol, 0.41 g KH_2PO_4, 1.21 g K_2HPO_4 in distilled water to make 500 ml of solution. Adjust pH to 7.2 with 1 N KOH. Store at 4° C.

100 ml 1% acidulated methyl green dispensed in five dropping bottles: dissolve 1.0 g methyl green in 99 ml distilled water + 1 ml glacial acetic acid.

40 ml 0.025% (w/v) Janus green B dispensed in five dropping bottles: dilute 1.0 ml of a 1% (w/v) aqueous stock solution of the stain with 39 ml of the mannitol grinding medium. The stock solution is stable if stored in a dark, stoppered bottle. The working solution should be prepared shortly before use.

PROJECT **9**

Students should work in teams of four; quantities are calculated for 20 students working in five teams.

Equipment

5 spectrophotometers	2 balances
1 refrigerated, high-speed centrifuge	2 hot plates

Supplies

10 mortars and pestles, chilled	5 15-ml test tubes
5 50-ml graduated cylinders	50 cuvettes
25 1-ml pipets	5 funnels
5 5-ml pipets	5 test tube racks
5 10-ml pipets	5 spatulas
5 pipet fillers, rubber bulb type	2 pairs scissors
5 single-edge razor blades	5 permanent marking pens
5 400-ml beakers for ice-water baths	Parafilm
5 French curve templates and rulers	Kimwipes
2 250-ml beakers for boiling-water baths	Pasteur pipets, 146 mm, and dropper bulbs

2 fresh, firm heads of cauliflower, refrigerated until used. Just before use, rinse with cold tap water and blot dry.

6 50-ml centrifuge tubes for first centrifugation

6 50-ml or 10 25-ml centrifuge tubes for second centrifugation

Cotton cheesecloth, grade 10 (Curity). Available from Thomas Scientific.

Reagents

25 g purified sea sand

500 ml grinding medium, 0.3 M D-mannitol–0.02 M phosphate buffer, pH 7.2: dissolve 27.33 g D-mannitol, 0.41 g KH_2PO_4, 1.21 g K_2HPO_4 in distilled water to make 500 ml of solution. Adjust pH to 7.2 with 1 N KOH. Store at 4° C.

500 ml assay medium, 0.3 M D-mannitol–0.02 M phosphate buffer–0.01 M KCl–0.005 M $MgCl_2$ dispensed in 10 Erlenmyer flasks: dissolve 27.33 g D-mannitol, 0.41 g KH_2PO_4, 1.21 g K_2HPO_4, 0.38 g KCl, 0.51 g $MgCl_2 \cdot 6H_2O$ in distilled water to make 500 ml of solution. Adjust pH to 7.2 with 1 N KOH. Store at 4° C.

50 ml 0.04 M sodium azide dispensed in five 15-ml test tubes: dissolve 0.26 g sodium azide* in distilled water to make 100 ml of solution.

*CAUTION: *Sodium azide is poisonous and in the dry state is explosive if subjected to shocks.*

50 ml 5 × 10⁻⁴ M DCIP dispensed in five 15-ml test tubes: dissolve 145 mg 2, 6-dichlorophenolindophenol sodium salt in distilled water to make 1000 ml of solution. Prepare shortly before use.

50 ml 0.2 M Na_2 malonate dispensed in five 15-ml test tubes: dissolve 3.32 g Na_2 malonate · H_2O (available from Eastman Kodak Co.) in distilled water to make 100 ml of solution. Adjust pH to 7.0 with 1 N HCl.

50 ml 0.2 M Na_2 succinate dispensed in five 15-ml test tubes: dissolve 5.40 g Na_2 succinate · $6H_2O$ in distilled water to make 100 ml of solution. Adjust pH to 7.0 with 1 N HCl.

PROJECT **10**

Students should work in pairs; quantities are calculated for 20 students working in 10 teams. Since the petroleum ether–acetone solvent is extremely flammable, the chromatography should be carried out in a fume hood.

Equipment

10 spectrophotometers. See note on page 251.

10 chromatography jars. Lids should have an affixed hook for attaching the chromatography paper. Small chromatography jars give better results than large jars.

4 balances

Supplies

Whatman 3MM chromatography paper

10 mortars and pestles, chilled

10 25-ml graduated cylinders

5 250-ml graduated cylinders

10 50-ml test tubes with corks

50 cuvettes with corks

10 French curve templates and rulers

10 small test tubes, 10 × 75 mm

Pasteur pipets, 146 mm, and dropper bulbs

10 capillary tubes, length 8–10 cm, i.d. 1 mm. Melting-point capillaries work well.

10 5-ml pipets

10 pipet fillers, rubber bulb type

100 g fresh spinach

10 test tube racks

10 pairs scissors

10 pairs forceps

10 permanent marking pens

Vaseline petroleum jelly

Kimwipes

Reagents

30 g purified sea sand

2 liters petroleum ether, 35°–60° C boiling range

1 liter acetone; 200 ml should be ice-cold.

PROJECT **11**

Students should work in pairs; quantities are calculated for 20 students working in 10 teams.

Equipment

10 spectrophotometers

4 balances

2 clinical centrifuges

Supplies

10	mortars and pestles, chilled	100	cuvettes
5	25-ml graduated cylinders	10	250-ml beakers
10	400-ml beakers for ice-water baths	10	thermometers
10	15-ml test tubes	10	funnels
50	1-ml pipets	10	test tube racks
20	5-ml pipets	10	pairs scissors
10	10-ml pipets	10	permanent marking pens
10	pipet fillers, rubber bulb type		Aluminum foil
10	French curve templates and rulers		Parafilm
20	15-ml conical centrifuge tubes		Kimwipes

200 g fresh spinach. Just before use, rinse with cold tap water and blot dry.

Pasteur pipets, 146 mm, and dropper bulbs

10 gooseneck or table lamps with 100-watt frosted incandescent bulbs

Cotton cheesecloth, grade 10 (Curity). Available from Thomas Scientific.

Reagents

30 g purified sea sand

500 ml 0.35 M NaCl–0.02 M Tris buffer, pH7.5, dispensed in 10 Erlenmyer flasks: dissolve 20.45 g NaCl, 2.42 g tris (hydroxymethyl) aminomethane in distilled water to make 1000 ml of solution. Adjust pH to 7.5 with 6 N HCl. Store at 4°C.

100 ml 4×10^{-4} M DCIP dispensed in 10 15-ml test tubes: dissolve 116 mg 2,6-dichlorophenolindophenol sodium salt in distilled water to make 1000 ml of solution. Prepare shortly before use.

100 ml 0.01 N ammonia dispensed in 10 15-ml test tubes: in a fume hood, prepare a 1 N solution by bringing 6.77 ml concentrated (28%) NH_4OH to 100 ml with distilled water. Then bring 5.0 ml of the 1 N solution to 500 ml with distilled water.

100 ml 10^{-4} M DCMU dispensed in 10 15-ml test tubes: dissolve 24 mg 3-(3,4-dichlorophenyl)-1, 1-dimethylurea in distilled water to make 1000 ml of solution. DCMU is available from K & K Labs, 121 Express St., Plainview, N.Y. 11803.

100 ml distilled water dispensed in 10 15-ml test tubes

PROJECT **12**

In the experiment on flagellar function, the deflagellation, centrifugation, and dispensing of samples should be done by one designated person.

Equipment

4	dissecting microscopes	1	clinical centrifuge
1	pH meter	1	reciprocating shaker
1	magnetic stirrer and stirring bar		
20	compound microscopes equipped with ocular micrometers		

Supplies

	Microscope slides	10	droppers
	Coverslips, 22 mm^2, No. 1½	20	test tube racks
80	small test tubes, 10 × 75 mm	20	permanent marking pens

1 100-ml beaker	Vaseline petroleum jelly
2 10-ml graduated cylinders	Toothpicks
2 15-ml graduated, conical centrifuge tubes	Kimwipes
2 50-ml Erlenmyer flasks with cotton plugs	

Pasteur pipets, 146 mm, and dropper bulbs

1 fluorescent lamp with two cool-white, 15-watt bulbs

Several cultures of *Pelomyxa carolinensis* and/or *Amoeba proteus*. *P. carolinensis* is preferred since they are larger. For short-term (7–10 days) maintenance of the cultures, add two previously boiled wheat grains to each culture as soon as they arrive from the biological supply company (assuming that no wheat or rice grains are already present in the culture). Long-term cultures can be maintained using the following medium: add three or four previously boiled wheat grains to 200 ml pasteurized (60° C for 15 min) spring water while still hot. Allow the medium to cool and then inoculate. For culturing *P. carolinensis*, also add *Paramecium caudatum*, on which the *P. carolinensis* feed.

Several cultures of *Paramecium caudatum*. For short-term (7–10 days) maintenance of the cultures, add two previously boiled wheat grains to each culture as soon as they arrive from the biological supply company (assuming that no wheat or rice grains are already present in the culture). Long-term cultures can be maintained using the following medium: add six to eight previously boiled wheat grains to 200 ml pasteurized (60° C for 15 min) spring water while still hot. Allow the medium to cool and then inoculate.

1 culture of *Chlamydomonas reinhardi*. For the experiment on flagellar function, the *Chlamydomonas* are cultured in Medium I of Sager and Granick (see Reagents), as follows. Using sterile technique, inoculate 75 ml of Medium I in a 250-ml Erlenmyer flask. The culture is grown at room temperature with constant aeration (cotton-filtered air) and constant illumination (fluorescent lamp). The culture is ready for use in the experiment when A_{675} is between 0.1 and 0.5, a culture density range that is reached 2–3 days after inoculation. Commercially available cultures of *Chlamydomonas* may be supplied growing on agar slants, on which the organisms lack flagella. Thus, even for the initial observations on *Chlamydomonas*, they must be grown in liquid Medium I for 2–3 days. Long-term cultures of *Chlamydomonas* are easily maintained on an agar slant of Medium I (10 ml Medium I + 0.15 g agar). The slant should be kept in a well-illuminated area at room temperature.

Reagents

10 ml suspension of Congo red–stained yeast: add 3 g compressed yeast and 30 mg Congo red to 10 ml distilled water; boil gently for 10 min.

100 ml Lugol's iodine solution dispensed in five dropping bottles: see page 247.

30 ml 0.5 N acetic acid dispensed in a dropping bottle: dilute 2.9 ml glacial acetic acid to 100 ml with distilled water.

30 ml 0.5 N KOH dispensed in a dropping bottle

150 ml Medium I of Sager and Granick: to a 1000-ml volumetric flask, add 0.5 g Na$_3$ citrate · 2H$_2$O, 0.1 g K$_2$HPO$_4$, 0.1 g KH$_2$PO$_4$, 0.3 g NH$_4$NO$_3$, 0.3 g MgSO$_4$ · 7H$_2$O, 0.04 g CaCl$_2$, 0.01 g FeCl$_3$ · 6H$_2$O, 10 ml trace metal solution (see below), distilled water to make 1000 ml of solution. Dissolve the sodium citrate first to avoid precipitation of any salts. The medium is autoclaved and then refrigerated until use. Medium I is also required for Project 13.

1000 ml trace metal solution: dissolve 100 mg H$_3$BO$_3$, 100 mg ZnSO$_4$ · 7H$_2$O, 40 mg MnSO$_4$ · 4H$_2$O, 20 mg CoCl$_2$ · 6H$_2$O, 20 mg Na$_2$MoO$_4$ · 2H$_2$O, 4 mg CuSO$_4$ in distilled water to make 1000 ml of solution.

5 bottles of Protoslo (Carolina Biological Supply Co.)

Students should work in teams of three; quantities are calculated for 21 sutdents working in seven teams. The deflagellation procedure and dispensing of samples should be done by one designated team.

Equipment

1 pH meter	1 clinical centrifuge
1 magnetic stirrer and stirring bar	1 reciprocating shaker
21 compound microscopes equipped with ocular micrometers	

Supplies

224 small test tubes, 10 × 75 mm	Microscope slides
1 50-ml graduated cylinder	Coverslips, 22 mm^2
4 10-ml pipets	1 100-ml beaker
1 pipet filler, rubber bulb type	21 test tube racks
3 15-ml graduated conical centrifuge tubes	21 permanent marking pens
	Kimwipes
4 50-ml Erlenmyer flasks with cotton plugs	
10 French curve templates and rulers	

Pasteur pipets, 146 mm, and dropper bulbs

1 fluorescent lamp with two cool-white, 15-watt bulbs

1 culture of *Chlamydomonas reinhardi*. The culture procedure is the same as in the experiment on flagellar function, Project 12, page 259.

Reagents

150 ml Medium I; see page 259.

10 ml Medium I containing colchicine, 3 mg/ml: dissolve 30 mg colchicine in 10 ml Medium I.

10 ml Medium I containing cycloheximide, 10 µg/ml: dissolve 10 mg cycloheximide in 10 ml Medium I. Then dilute 0.5 ml of this solution to 50 ml with Medium I.

100 ml Lugol's iodine solution dispensed in five dropping bottles; see page 247.

30 ml 0.5 N acetic acid dispensed in a dropping bottle; see page 259.

30 ml 0.5 N KOH dispensed in a dropping bottle

Optional Exercises

1. To determine whether the effects of colchicine are reversible, the colchicine medium can be replaced with fresh Medium I (without colchicine) part way through the experiment. After the 20-min sampling, the colchicine culture is centrifuged for 5 min at 1300 g and the pellet resuspended in 10 ml fresh Medium I. The suspension is transferred to a clean 50-ml Erlenmyer flask, returned to the shaker and sampled for the duration of the experiment. For this study, the initial concentration of colchicine should be 1.5 mg/ml.

2. To ascertain whether colchicine induces disassembly of flagellar microtubules, the colchicine can be added after some regeneration has occurred. Specifically, an aliquot of the deflagellated culture is allowed to undergo regeneration in Medium I for 30 min (just like the control culture). After the 30-min sampling, colchicine is added directly to the culture (final concentration = 3 mg/ml). Flagellar length is measured at the same sampling times used for the other cultures.

PROJECT 14

For processing the root tips for the Feulgen reaction, students should work in teams of four, but for making the squashes, students should work individually. Quantities are calculated for 20 students working in five teams. If the mitotic stages are to be photographed in a future lab session, the squash preparations should be made permanent.

Students should be provided with fixed root tips that are ready to be stained. *Vicia faba* seeds need to be germinated about 2 weeks before the project is to be done. Soak the seeds in a 10% aqueous solution of Clorox (or comparable hypochlorite product) for 10 minutes, rinse well in running water, and then plant in sterile vermiculite. Water the vermiculite initially and every few days thereafter. When the primary root is 5 to 8 cm long (about a week after planting) cut off the root tip and replace the seedling in the vermiculite. In about a week there should be many lateral roots. The terminal 4 to 7 mm of the lateral roots should be fixed for 24 hours in freshly prepared 1:3 acetic:ethanol (1 vol glacial acetic acid:3 vol absolute ethanol) and then transferred to 70% ethanol, in which they can be stored almost indefinitely at 4° C.

As an alternative to having students prepare the root tip squashes, prepared slides can be supplied. Feulgen-stained squashes of *V. faba* root tips are available from Carolina Biological Supply Co. (cat no. B555V) and Ward's Natural Science Establishment, Inc. (cat. no. 93W1910).

Equipment
20 compound microscopes   2 60° C water baths

Supplies
Microscope slides with frosted end   5 50-ml beakers
Coverslips, 22 mm^2 20 pairs forceps
20 single-edge razor blades   Kimwipes
20 glass rods with fire-polished, flat ends
40 fixed root tips of *Vicia faba*. Seeds can be purchased from W. Atlee Burpee Co., 300 Park Avenue, Warminster, Pa. 18991.
Cotton cheesecloth, grade 10 (Curity). Available from Thomas Scientific.
Vaseline petroleum jelly and toothpicks. (These items are not required if permanent squash preparations will be made.)

Reagents
100 ml 1 N HCl dispensed in five 50-ml beakers; see page 249.
100 ml Schiff's reagent dispensed in five capped vials; see page 249.
100 ml 45% (v/v) glacial acetic acid dispensed in five dropping bottles

For Permanent Squash Preparations
Dry ice
320 ml 95% ethanol dispensed in eight Coplin jars
320 ml absolute ethanol dispensed in eight Coplin jars
80 ml 0.5% (w/v) fast green FCF in 95% ethanol, dispensed in two Coplin jars
50 ml Euparal dispensed in five glass-rod dropping bottles

PROJECT 15

Two laboratory sessions are often required for identification of all the stages of meiosis. Students should be provided with fixed, stained testes from last instar or young adult males. A species commonly used for chromosome analysis in North America is *Melanoplus femur-rubrum*, although many other species are also suitable. Grasshoppers can be collected locally, with one collection trip providing enough material for several years' use. An excellent pictorial key for specimen identification is *How to Know the Grasshoppers, Cockroaches and their Allies* by J. R. Helfer (Wm C. Brown Co., Dubuque, Iowa, 1963).

In grasshoppers, the testes are fused into one large, yellow-orange body, which is exposed by tearing open the ventral abdomen with a pointed forceps. After removal, it is immediately fixed in freshly prepared 1:3 acetic:ethanol (1 vol glacial acetic acid: 3 vol absolute ethanol). The dissection and fixation can be done in the field. Following overnight fixation, the tissue is transferred to 70% ethanol, in which it can be stored almost indefinitely at 4° C. The entire testis is stained by placing it in alcoholic HCl-carmine (see Reagents) for 16 hr. The stained testis can also be stored in 70% ethanol at 4° C. If the meiotic stages are to be photographed in a future lab session, the squash preparations should be made permanent.

As an alternative to having students prepare the testis squashes, prepared slides can be supplied. Aceto-carmine squashes of grasshopper testis are available from Ward's Natural Science Establishment, Inc. (cat nos. 93W2246 and 93W2249).

Equipment

20 dissecting microscopes	20 compound microscopes

Supplies

Microscope slides	Coverslips, 22 mm^2
20 pairs dissecting needles	40 Syracuse dishes
10 fixed, stained grasshopper testes	Kimwipes

Vaseline petroleum jelly and toothpicks. (These items are not required if permanent squash preparations will be made.)

Reagents

100 ml alcoholic HCl-carmine (method of R. Snow): dissolve 4.0 g carmine alum lake (Fisher Scientific) in 10 ml 1 N HCl + 20 ml distilled water in a 125-ml Erlenmyer flask. Plug the flask with cotton and boil gently for 10 min. When cool, add 70 ml absolute ethanol and then filter through coarse filter paper with suction. The stain can be reused and lasts almost indefinitely if stored in a dark, tightly stoppered bottle. It does not need to be filtered again.

200 ml 70% ethanol

200 ml 45% (v/v) glacial acetic acid dispensed in 10 dropping bottles

For Permanent Squash Preparations

Supply the materials listed on page 261, as well as 20 pairs forceps and 20 single-edge razor blades. The fast green is not needed.

PROJECT **16**

For the silver staining, students should be provided with unstained slide preparations from short-term human leukocyte cultures harvested any time up to two weeks prior to the lab session in which the silver staining will be done. A total of three cultures should suffice for a class of 20 students since each harvested culture yields 9 to 10 slides. For all cultures, the blood should come from a healthy individual other than a student, and no attempt should be made to karyotype the cells.

The instructions for initiating and incubating the blood cultures and for harvesting the cells follow. Care should be taken when handling blood since there is always the slight possibility that the blood is contaminated with infectious agents.

Accordingly, disposable gloves should be worn, spills wiped up, the area disinfected, and waste materials disposed of properly. It is advisable that the individual doing the harvest be the one who initiates the culture with his/her own blood.

1. Reconstitute the lyophilized chromosome medium with the supplied reconstituting fluid and place in a 37° C incubator for 15 min.

2. Use aseptic technique to obtain the blood sample. After washing the hands with soap and water, and drying them, clean the tip of the ring or middle finger with a sterile alcohol swab. When the fingertip has dried, lance it with a sterile lancet, "milk" the finger, and collect three heparinized capillary tubes of blood (four to five drops). Each capillary tube is filled by holding it nearly level and touching the tip to the drop. Remember to clean the lanced fingertip with alcohol and to discard the lancet, which should never be used more than once.

3. Using a capillary dropper bulb, transfer the blood to the reconstituted medium. Invert the capped culture several times to mix the contents. Record the date and time that the culture was started.

4. Incubate the culture at 37° C for 66–72 hr. Twice each day, invert the culture several times to dislodge the cells from the bottom and disperse the culture.

5. Two hr before the incubation is finished, add 0.4 ml of the colchicine solution* to the culture. Invert the capped culture several times before returning it to the incubator.

6. After the 2-hr incubation with colchicine, disperse the culture by inverting the vial several times. Then, with a Pasteur pipet, transfer the entire culture to a 15-ml conical centrifuge tube. Centrifuge at 500 g for 5 min.

7. With the Pasteur pipet, carefully aspirate and discard almost all of the liquid above the pellet. Add 5 ml of warm (37° C) Hanks balanced salt solution, resuspend the cells with the Pasteur pipet, and centrifuge at 500 g for 5 min. Whenever resuspending, continue until all clumps are dispersed.

8. With the Pasteur pipet, carefully aspirate and discard almost all of the liquid above the pellet. Add 5 ml of 0.075 M KCl (room temperature) and gently resuspend the cells with the Pasteur pipet. Allow the cells to remain in the hypotonic solution for 12 min.

9. Centrifuge at 200 g for 5 min. While the cells are being centrifuged, prepare the fixative as follows. Add 45 ml methanol and 15 ml glacial acetic acid to an Erlenmyer flask. Mix well and keep the flask capped.

10. With the Pasteur pipet, carefully aspirate and discard almost all of the liquid above the pellet. Then, carefully draw up all of the pellet (cells) into the lower, tapered portion of the Pasteur pipet. Place the pipet aside.

11. Add 5 ml of fixative† to the nearly empty centrifuge tube. Place the tip of the Pasteur pipet (containing the cells) just beneath the surface of the fixative and slowly compress the pipet bulb so that the cells run down the side of the centrifuge tube and settle at the bottom. The dark-brown color is due to the action of the fixative on the hemoglobin released from the hemolyzed erythrocytes. The pellet will progressively whiten with subsequent changes of the fixative.

12. Gently resuspend the cells with the Pasteur pipet. Cover the tube with Parafilm and allow the cells to fix for 15 min. The swollen, hypotonic-treated cells are rather fragile and should be handled very gently.

13. Centrifuge at 200 g for 5 min.

14. With the Pasteur pipet, carefully aspirate and discard almost all of the fixative above the pellet.

*CAUTION: *Colchicine is poisonous; do not pipet by mouth.*
†CAUTION: *The fixative is poisonous; do not pipet by mouth.*

15. Add 4 ml of fresh fixative and gently resuspend the cells with the Pasteur pipet.
16. Centrifuge, aspirate, and resuspend the pellet in 4 ml of fresh fixative, as described in steps 13, 14, and 15.
17. Again, centrifuge, aspirate, and resuspend as described in steps 13, 14, and 15, but this time gently resuspend the pellet in only 0.3–0.4 ml of fresh fixative.
18. Make the air-dried preparations on alcohol-cleaned slides that have been stored in chilled, distilled water in Coplin jars. Wipe the water from the bottom of the slide and then with the Pasteur pipet held 6–10 cm above the slide, let two or three drops of the cell suspension drop onto the horizontal, wet slide surface. Blow vigorously on the surface and immediately place the slide on a hot plate until almost dry (about 10 sec). The temperature of the hot place should be set to maintain the temperature of a small beaker of water at 50°-55° C.
19. Store the unstained slide preparations in a lightproof slide box.

For the silver staining, each student should work individually, but reagents should be shared by each pair of students. The items required for the silver staining are listed separately from the items required for the cell culture and harvest, and quantities are calculated for 20 students working in 10 teams. Since the silver solutions stain skin and clothing, it is imperative that lab coats and disposable gloves be worn. As an added precaution, the lab bench and floor beneath it should be covered (with Benchkote and a drop cloth, for example).

Equipment for Cell Culture and Harvest

1 37° C incubator   1 clinical centrifuge

1 hot plate set to maintain temperature of a 50-ml beaker of water at 50°-55° C

Supplies for Cell Culture and Harvest

1 50-ml graduated cylinder	Disposable gloves
1 1-ml pipet	Sterile alcohol swabs
2 5-ml pipets	1 pair forceps
1 pipet filler, rubber bulb type	Parafilm
1 test tube rack	Kimwipes

Pasteur pipets, 146 mm, and dropper bulbs

3 15-ml graduated, concial centrifuge tubes

2 125-ml Erlenmyer flasks with screw caps

Microscope slides with frosted end. New slides from the box should be cleaned with 95% ethanol and stored in cold distilled water in Coplin jars.

20 Giemsa-stained slide preparations of human chromosomes. They can be purchased from Carolina Biological Supply Co. or Ward's Natural Science Establishment, Inc. Alternatively, they can be prepared by harvesting two or three additional cultures and staining the slides with Giemsa blood stain (see p. 227, steps 7 and 8).

Reagents for Cell Culture and Harvest

2 Difco TC Chromosome Microtest Kits. Each kit contains material for two cultures: two vials each of lyophilized culture medium, reconstituting fluid, arresting solution (colchicine), and Hanks balanced salt solution. Each kit also contains sterile blood lancets, Pasteur pipets with bulbs, sterile capillary tubes (for collecting blood) with bulbs, and Giemsa stain. (Difco Laboratories, P.O. Box 1058, Detroit, Michigan 48232, cat. no. 5060-32-8.)

50 ml 0.075 M KCl: dissolve 5.59 g KCl in distilled water to make 1000 ml of solution

45 ml reagent grade methanol from an unopened bottle

15 ml reagent grade glacial acetic acid from an unopened bottle

Equipment for Silver Staining
1 50°–60° C oven   20 compound microscopes

Supplies for Silver Staining
40 Coplin jars for "stop" rinses   Benchkote
20 Petri dishes   Drop cloths
20 pairs forceps   Disposable gloves
    Coverslips, 22 × 50 mm   Parafilm
10 250-ml beakers for ice-water baths
    Pasteur pipets, 146 mm, and dropper bulbs
    Slide holders to prop slides in a near vertical position. Available from Thomas
        Scientific.

Reagents for Silver Staining
30 ml 50% silver nitrate solution dispensed in 10 small (10 × 75 mm) test tubes:
    dissolve 15.0 g silver nitrate in distilled-deionized water to make 30 ml of
    solution. Just before use, filter the solution through a fresh 0.20 μm filter
    (Nalgene Filter Unit, cat. no. 120-0020). The solution is used at room
    temperature but is stored at 4° C in a dark stoppered bottle.

20 ml ammoniacal silver (AS) solution dispensed in 10 small (10 × 75 mm) test
    tubes: in a fume hood, dissolve 8.0 g silver nitrate in 10 ml distilled-deionized
    water + 10 ml concentrated (28% w/w) NH_4OH. Just before use, filter the
    solution through a fresh 0.20 μm filter. The AS solution is kept in an ice-water
    bath during use and is stored at 4° C in a dark stoppered bottle.

20 ml formalin developer dispensed in 10 small (10 × 75 mm) test tubes: mix 3.0
    ml formalin (37% w/w formaldehyde) with 97 ml distilled-deionized water.
    Before use, neutralize the solution (pH 7.0) by adding sodium acetate crystals
    and then adjust the pH to 4.5 with formic acid. This solution is kept in an
    ice-water bath during use and is stored at 4° C.

175 ml xylene dispensed in five Coplin jars

50 ml Permount dispensed in five glass-rod dropping bottles

PROJECT **17**

For the Hoechst–Giemsa staining, students should be provided with unstained slide
preparations from short-term human leukocyte cultures harvested up to one week
prior to the lab session in which the Hoechst-Giemsa staining will be done. A total of
three cultures should suffice for a class of 20 students since each harvested culture
yields 9 to 10 slides. For all cultures, the blood should come from a healthy individual
other than a student, and no attempt should be made to karyotype the cells.

The instructions for initiating and incubating the blood cultures and for
harvesting the cells are given on pages 263–264; the two additional steps listed here
are also required for this project. Care should be taken when handling blood since
there is always the slight possibility that the blood is contaminated with infectious
agents. Accordingly, disposable gloves should be worn, spills wiped up, the area
disinfected, and waste materials disposed of properly. It is advisable that the
individual doing the harvest be the one who initiates the culture with his/her own
blood.

1. Wrap a piece of aluminum foil around the culture vials so that they will not be
   exposed to light during the incubation period.

2. Add 0.1 ml of the BrdU solution* to each reconstituted culture prior to adding the blood.

The equipment, supplies, and reagents required for initiating and harvesting three leukocyte cultures are listed on pages 264–265. Required items for the BrdU–Hoechst–Giemsa procedure are listed below.

Equipment and Supplies

20 compound microscopes	Coverslips, 22 × 50 mm
2 2-ml pipets	10 Coplin jars
2 10-ml pipets	20 pairs forceps
2 pipet fillers, rubber bulb type	Aluminum foil
2 50-ml graduated cylinders	Benchkote

3 fluorescent lamps, each with two cool-white bulbs

Pasteur pipets, 146 mm, and dropper bulbs

Slide holders to prop slides in a near vertical position. Available from Thomas Scientific.

Reagents.

31 ml 2 × 10^{-3} M BrdU in Hanks balanced salt solution: dissolve 19 mg 5-bromo-2'-deoxyuridine† (Sigma Chemical Co.) in 31 ml sterile Hanks solution (Difco Laboratories). The solution can be filter-sterilized or used as is, since there are antibiotics in the culture medium. The addition of 0.1 ml to a 4-ml culture will give a final BrdU concentration of 5 × 10^{-5} M.

160 ml aqueous solution of Hoechst 33258, 200 μg/ml, dispensed in four Coplin jars: dissolve 32 mg Hoechst 33258‡ in 160 ml distilled water. The dye is available from Sigma Chemical Co., Catalog No. B2883.

250 ml McIlvaine's buffer, pH 7.0: dissolve 0.85 g citric acid, anhydrous (0.93 g citric acid, monohydrate), 5.85 g Na_2HPO_4 in distilled water to make 250 ml of solution.

100 ml M/15 phosphate buffer, pH 6.8: dissolve 4.73 g Na_2HPO_4, 4.54 g KH_2PO_4 in distilled water to make 1000 ml of solution.

25 ml Harleco Giemsa Blood stain, Original Azure Blend Type, available from American Scientific Products.

175 ml xylene dispensed in five Coplin jars

50 ml Permount dispensed in five glass-rod dropping bottles

PROJECT **18**

The following preparations are recommended for photomicrography: Wright- or Giemsa-stained blood cells, mitotic stages in root-tip squashes of *Vicia faba*, and meiotic stages in testis squashes of the grasshopper.

Listed are the basic items required for photomicrography, film processing, and printing. A darkroom is required only for printing; Technical Pan Film 2415 can be loaded in a developing tank using a changing bag. For the latest, most complete information on any Kodak or Polaroid product, always refer to the manufacturer's instructions. For technical assistance, there are also toll-free numbers for Kodak (800-242-2424) and Polaroid (800-343-5000).

---

*CAUTION: *BrdU is toxic; do not pipet by mouth.*

†CAUTION: *Wear gloves and mask when weighing out BrdU.*

‡CAUTION: *Wear gloves and mask when weighing out Hoechst 33258.*

Equipment

Trinocular microscopes equipped with 35mm and Polaroid attachment cameras, and with exposure meters

Enlarger

Polaroid 35mm AutoProcessor

Supplies

Stage micrometer slide	Photographic timer
Kodak Projection Print Scale	Print tongs
Scissors	Squeegee
Film clips	Paper cutter
Glassine envelopes	Tacking iron
Polaroid 35mm Slide Mounter	500-ml graduated cylinder
Polaroid 35mm Slide Mounts	600-ml beaker
Enlarger easel	Funnels
Enlarger focusing scope	Thermometers
Camel's hair brush	Changing bag (optional)

Kodak Dry Mounting Tissue, Type 2

Polycontrast filters: PC 2, PC 2½, PC 3, PC 3½, PC 4

Kodak Technical Pan Film 2415. Available in magazines with 36 exposures, as well as in 150-foot rolls for economical bulk loading. Unexposed film can be stored almost indefinitely in the refrigerator or freezer. Allow 2–3 hr for refrigerated film to reach room temperature before opening the package; for frozen film, allow 5 hr.

Bulk film loader and reusable 35mm magazines (optional)

Kodak Polycontrast Rapid II RC Paper, F (glossy), medium weight. Unexposed print paper can be stored almost indefinitely in the refrigerator or freezer. Allow it to come up to room temperature before opening the package.

Polaroid Type 667 (8 exposures) and Polapan (36 exposures). Store in refrigerator and allow 2–3 hr for the film to reach room temperature before opening the package. *Do not freeze.* Each roll of Polapan comes complete with a processing pack.

Filters: Assorted neutral density filters (50%, 25%, 12.5% transmittance); Wratten filter No. 58 and/or No. 11. Wratten gelatin filters are easier to handle and more durable if cut and then mounted in 2-in. × 2-in. glass projection slides.

Plastic 35mm developing tank (Paterson or Star-D, for example)

Safelight with 15-watt bulb and OC filter

Trays for developer, stop bath, and fixer

Reagents

All the photgraphic chemicals listed are toxic if swallowed or inhaled, and repeated contact may cause skin irritation and allergic skin reaction.

Kodak HC-110 developer. Make up the stock solution according to package directions and then dilute just before use. Dilution F (for moderate contrast) is given in the project. If high-contrast Technical Pan negatives are desired, use dilution D (1:10): mix 30 ml HC-110 stock solution + 270 ml water; development time is 6 min at 20° C. With dilution D, Technical Pan has an ASA of 100.

Kodak Indicator Stop Bath. It can be used for many rolls of film and many prints until the solution turns blue.

Kodak Dektol developer. One gallon has a capacity for 100 8-in. × 10-in. prints.
Kodak Fixer. One gallon has a capacity for 100 8-in. × 10-in. prints.
Kodak Photo Flo, diluted as per package directions
Kodak Hypo Clearing Agent (optional)